AIR, WATER AND SOIL POLLUTION SCIENCE AND TECHNOLOGY SERIES

POISONING AND ACIDIFICATION OF THE EARTH'S OCEANS

AIR, WATER AND SOIL POLLUTION SCIENCE AND TECHNOLOGY SERIES

Trends in Air Pollution Research
James, V. Livingston (Editor)
2005. ISBN: 1-59454-326-7

Agriculture and Soil Pollution: New Research
James, V. Livingston (Editor)
2005. ISBN: 1-59454-310-0

Water Pollution: New Research
A.R. Burk (Editor)
2008. ISBN: 1-59454-393-3

Air Pollution: New Research
James, V. Livingston (Editor)
2007. ISBN: 1-59454-569-3

Air Pollution Research Advances
Corin G. Bodine (Editor)
2007. ISBN: 1-60021-806-7

Marine Pollution: New Research
Tobias N. Hofer (Editor)
2008. ISBN: 978-1-60456-242-2

Complementary Approaches for Using Ecotoxicity Data in Soil Pollution Evaluation
M. D. Fernandez and J. V. Tarazona
2008. ISBN: 978-1-60692-105-0

Complementary Approaches for Using Ecotoxicity Data in Soil Pollution Evaluation
M. D. Fernandez and J. V. Tarazona
2008. ISBN: 978-1-60876-411-2 (Online Book)

Lake Pollution Research Progress
Franko R. Miranda and Luc M. Bernard (Editors)
2008. ISBN: 978-1-60692-106-7

Lake Pollution Research Progress
Franko R. Miranda and Luc M. Bernard (Editors)
2008. ISBN: 978-1-60741-905-1 (Online Book)

River Pollution Research Progress
Mattia N. Gallo and Marco H. Ferrari (Editors)
2009. ISBN: 978-1-60456-643-7

Heavy Metal Pollution
Samuel E. Brown and William C. Welton (Editors)
2008. ISBN: 978-1-60456-899-8

Heavy Metal Pollution
Samuel E. Brown and William C. Welton (Editors)
2008. ISBN: 978-1-61668-049-7 (Online book)

Cruise Ship Pollution
Oliver G. Krenshaw (Editor)
2009. ISBN: 978-1-60692-655-0

Water Purification
Nikolaj Gertsen and Linus Sønderby (Editors)
2009. ISBN: 978-1-60741-599-2

Water Purification
Nikolaj Gertsen and Linus Sønderby (Editors)
2009. ISBN: 978-1-61668-275-0 (Online book)

Environmental and Regional Air Pollution
Dean Gallo and Richard Mancini (Editors)
2009. ISBN: 978-1-60692-893-6

Environmental and Regional Air Pollution
Dean Gallo and Richard Mancini (Editors)
2009. ISBN: 978-1-60876-553-9 (Online Book)

Industrial Pollution including Oil Spills
Harry Newbury and William De Lorne (Editors)
2009. ISBN: 978-1-60456-917-9

Traffic Related Air Pollution and Internal Combustion Engines
Sergey Demidov and Jacques Bonnet (Editors)
2009. ISBN: 978-1-60741-145-1

Sludge: Types, Treatment Processes and Disposal
Richard E. Baily (Editor)
2009. ISBN: 978-1-60741-842-9

National Air Quality: Status and Trends
Joseph DiSano *(Editor)*
2010. ISBN: 978-1-60741-513-8

Poisoning and Acidification of the Earth's Oceans
Geoffrey E. Mason *(Editor)*
2010. ISBN: 978-1-60741-560-2

Sulphate-Reducing Bacteria in Biological Treatment Wastewaters
Dorota Wolicka
2010. ISBN: 978-1-60876-931-5

Pollution Control: Management, Technology and Regulations
Horatio R. Velasquez
2010. ISBN: 978-1-61668-584-3

Air Pollution and Ship Emissions
Jacob Boutin *(Editor)*
2010. ISBN: 978-1-60876-087-9

Heavy Metal Compounds in Soil: Transformation upon Soil Pollution and Ecological Significance
Tatiana M. Minkina, Galina V. Motusova, Olga G. Nazarenko and Saglara S. Mandzhieva
2010. ISBN: 978-1-60876-466-2

From Soil Contamination to Land Restoration
Claudio Bini
2010. ISBN: 978-1-60876-853-0

Green Plants and Pollution
Rajiv Sinha
2010. ISBN: 978-1-61668-147-0

Green Plants and Pollution
Rajiv Sinha
2010. ISBN: 978-1-61668-403-7 (Online book)

AIR, WATER AND SOIL POLLUTION SCIENCE AND TECHNOLOGY SERIES

POISONING AND ACIDIFICATION OF THE EARTH'S OCEANS

GEOFFREY E. MASON
EDITOR

Nova Science Publishers, Inc.
New York

For permission to use material from this book please contact us:
Telephone 631-231-7269; Fax 631-231-8175
Web Site: http://www.novapublishers.com

LIBRARY OF CONGRESS CATALOGING-IN-PUBLICATION DATA

Poisoning and acidification of the Earth's oceans / editor, Geoffrey E. Mason.
p. cm.
Includes index.
ISBN 978-1-60741-560-2 (hardcover)
1. Seawater--Carbon dioxide content--Environmental aspects. 2. Marine animals--Effects of water acidification on. 3. Ocean-atmosphere interaction. 4. Chemical weapons disposal--United States. I. Mason, Geoffrey, 1948-
GC117.C37P65 2009
363.739'4--dc22

2009041943

Published by Nova Science Publishers, Inc. ✢ New York

CONTENTS

PREFACE

With increasing concentrations of carbon dioxide (CO2) in the atmosphere, the extent of effects on the ocean and marine resources is an increasing concern. One aspect of this issue is the potential for seawater to become increasing acidic (i.e., ocean acidification). Scientists are concerned that increasing acidity could result in reduced growth or even death of shell-forming animals as well as disruption of marine food webs and reproductive physiology. Congress is beginning to focus attention on better understanding this issue and determining how it might be addressed. As well as examining the above issues, this book also presents an overview of the causes of hypoxia, the U.S. areas of most concern, federal legislation, and relevant research programs. In addition, this book reviews the history and oil spills in the United States; identifies the legal authorities governing oil spill prevention, response and cleanup; and examines the threats of future oil spills in U.S. coastal waters.

This book consists of public documents which have been located, gathered, combined, reformatted, and enhanced with a subject index, selectively edited and bound to provide easy access.

Chapter 1 - Good morning Chairman Cantwell and members of the Subcommittee. Thank you for giving me the opportunity to speak with you today on the short- and long-term impacts of ocean acidification on marine resources. My name is Richard Feely, I am a Supervisory Chemical Oceanographer at the Pacific Marine Environmental Laboratory of the National Oceanic and Atmospheric Administration (NOAA) in Seattle, WA. My personal area of research is the study of the oceanic carbon cycle and its impact on marine organisms. I have worked for NOAA for more than 32 years and have published more than 160 peer-reviewed scientific journal articles, book chapters and technical reports. I serve on the U.S. Carbon Cycle Science Program Scientific

Steering Group and I am the co-chair of the U.S. Repeat Hydrography Program Scientific Oversight Committee. For today, you have asked me to provide my insights on ocean acidification and its effect on living marine ecosystems. Most of my comments below are derived from the Royal Society Report, *"Ocean Acidification Due to Increasing Atmospheric Carbon Dioxide"* (Raven *et al.*, 2005) and the recent U.S. report, derived from a workshop held jointly by the National Science Foundation (NSF), NOAA, and the U.S. Geological Survey, entitled *"Impacts of Ocean Acidification on Coral Reefs and Other Marine Calcifiers"* (Kleypas *et al.*, 2006).

Chapter 2 - An adequate level of dissolved oxygen is necessary to support most forms of aquatic life. While very low levels of dissolved oxygen (hypoxia) can be natural, especially in deep ocean basins and fjords, hypoxia in coastal waters is mostly the result of human activities that have modified landscapes or increased nutrients entering these waters. Hypoxic areas are more widespread during the summer, when algal blooms stimulated by spring runoff decompose to diminish oxygen. Such hypoxic areas may drive out or kill animal life, and usually dissipate by winter. In many places where hypoxia has occurred previously, it is now more severe and longer lasting; in others where hypoxia did not exist historically, it now does, and these areas are becoming more prevalent.

The largest hypoxic area affecting the United States is in the northern Gulf of Mexico near the mouth of the Mississippi River, but there are others as well. Most U.S. coastal estuaries and many developed nearshore areas suffer from varying degrees of hypoxia, causing various environmental damages. Research has been conducted to better identify the human activities that affect the intensity and duration of, as well as the area affected by, hypoxic events, and to begin formulating control strategies.

Near the end of the 105th Congress, the Harmful Algal Bloom and Hypoxia Research and Control Act of 1998 was signed into law as Title VI of P.L. 105-383. Provisions of this act authorize appropriations through NOAA for research, monitoring, education, and management activities to prevent, reduce, and control hypoxia. Under this legislation, an integrated Gulf of Mexico hypoxia assessment was completed in the late 1990s. In 2004, Title I of P.L. 108-456, the Harmful Algal Bloom and Hypoxia Amendments Act of 2004, expanded this authority and reauthorized appropriations through FY2008. Legislation has been introduced in the 110th Congress to reauthorize and amend the Harmful Algal Bloom and Hypoxia Research and Control Act.

As knowledge and understanding have increased concerning the possible impacts of hypoxia, congressional interest in monitoringand addressing the problem has grown. The issue of hypoxia is seen as a search for (1) increased scientific knowledge and understanding of the phenomenon, as well as (2) cost-effective actions that might diminish the size of hypoxic areas by changing practices that promote their growth and development. This report presents an overview of the causes of hypoxia, the U.S. areas of most concern, federal legislation, and relevant federal research programs. This report will be updated as circumstances warrant.

Chapter 3 - With increasing concentrations of carbon dioxide (CO_2) in the atmosphere, the extent of effects on the ocean and marine resources is an increasing concern. One aspect of this issue is the potential for seawater to become increasing acidic (i.e., ocean acidification). Scientists are concerned that increasing acidity could result in reduced growth or even death of shell-forming animals (e.g., molluscs and certain planktonic organisms) as well as disruption of marine food webs and reproductive physiology. Congress is beginning to focus attention on better understanding this issue and determining how it might be addressed.

Chapter 4 - The Marine Protection, Research, and Sanctuaries Act has two basic aims: to regulate intentional ocean disposal of materials, and to authorize related research. Permit and enforcement provisions of the law are often referred to as the Ocean Dumping Act. The basic provisions of the act have remained virtually unchanged since 1972, when it was enacted to establish a comprehensive waste management system to regulate disposal or dumping of all materials into marine waters that are within U.S. jurisdiction, although a number of new authorities have been added. This report presents a summary of the law.

During the past two decades, while U.S. oil imports and consumption have steadily risen, oil spill incidents and the volume of oil spilled have not followed a similar course. In general, the annual number and volume of oil spills have shown declines — in some cases, dramatic declines. The 1989 *Exxon Valdez* spill in Alaskan waters played a large role in stimulating actions that contributed to this trend, particularly the decrease in the annual spill volumes. The *Exxon Valdez* spill highlighted the need for stronger legislation, inflamed public sentiment, and spurred Congress to enact comprehensive oil spill legislation, resulting in the Oil Pollution Act of 1990 (P.L. 101-380). This law expanded and clarified the authority of the federal government and created new oil spill prevention and preparedness requirements. Moreover, the 1990 legislation strengthened existing liability provisions, providing a greater deterrent against spills. After 1990, spill volume from oil tankers, the vessels that carry and have spilled the most oil, decreased significantly.

Considering that U.S. oil consumption and oil imports have steadily increased, the trend of declining spill incidents and volume in past years is noteworthy. Yet, recent annual data indicate that the overall decline of annual spill events may have stopped. Although Energy Information Administration projections indicate that oil imports are expected to decline in coming years, the United States is projected to continue importing more than 50% of the oil it consumes. The threat of oil spills raises the question of whether U.S. officials have the necessary resources at hand to respond to a major spill. There is some concern that the favorable U.S. spill record has resulted in a loss of experienced personnel, capable of responding quickly and effectively to a major oil spill. Moreover, the level of funding required to respond to such a spill, particularly its aftermath, may be currently inadequate, according to U.S. Coast Guard reports.

Recent oil spills into San Francisco Bay (November 2007) and the Mississippi River (July 2008) have generated further interest and debate concerning oil spill policy issues, such as oil spill response and prevention.

No oil spill is entirely benign. Even a relatively minor spill, depending on the timing and location, can cause significant harm to individual organisms and entire populations. Marine mammals, birds, bottom-dwelling and intertidal species, and organisms in early developmental stages — eggs or larvae — are especially vulnerable to a nearby spill. However, the effects of oil spills can vary greatly. Oil spills can cause impacts over a range of time scales, from only a few days to several years, or even decades in some cases.

This report reviews the history and trends of oil spills in the United States; identifies the legal authorities governing oil spill prevention, response, and cleanup; and examines the threats of future oil spills in U.S. coastal waters.

The U.S. Armed Forces disposed of chemical weapons in the ocean from World War I through 1970. At that time, it was thought that the vastness of ocean waters would absorb chemical agents that may leak from these weapons. However, public concerns about human health and environmental risks, and the economic effects of potential damage to marine resources, led to a statutory prohibition on the disposal of chemical weapons in the ocean in 1972. For many years, there was little attention to weapons that had been dumped offshore prior to this prohibition. However, the U.S. Army completed a report in 2001 indicating that the past disposal of chemical weapons in the ocean had been more common and widespread geographically than previously acknowledged. The Army cataloged 74 instances of disposal through 1970, including 32 instances off U.S. shores and 42 instances off foreign shores. The disclosure of these records has renewed public concern about lingering risks from chemical weapons still in the ocean today.

The risk of exposure to chemical weapons dumped in the ocean depends on many factors, such as the extent to which chemical agents may have leaked into seawater and been diluted or degraded over time. Public health advocates have questioned whether contaminated seawater may contribute to certain symptoms among coastal populations, and environmental advocates have questioned whether leaked chemical agents may have affected fish stocks and other marine life. There also has been public concern that chemical weapons could wash ashore or be accidentally retrieved during activities that disturb the seabed, such as dredging and trawl fishing. Although such incidents have occurred domestically and abroad, they are rare relative to the thousands of weapons dumped in the ocean. Assessing the degree of risks is difficult because of a lack of information.

Whether the risks are low or high, how to respond to them is fraught with many challenges. The primary obstacle is locating the weapons in the ocean. The lack of coordinates for most of the disposal sites, and the possibility that ocean currents may have moved weapons beyond these areas, makes finding the weapons difficult at best, if not impracticable in some cases. Enacted in the second session of the 109th Congress, the John Warner National Defense Authorization Act for FY2007 (P.L. 109-364, H.R. 5122) requires further review of historical records to attempt to identify where chemical and conventional weapons were dumped off U.S. shores, research of the effects of these weapons on the ocean environment, and monitoring to determine whether contamination or health or safety risks are present. The funding and implementation of these requirements are potential issues for the 110th Congress.

In the event that the weapons are located, retrieving them from the seabed could be technically challenging and could introduce new risks during retrieval and transport for onshore disposal. Leaving located weapons in place, and warning the public to avoid these areas, may be more feasible and involve fewer immediate risks. However, long-term risks would remain. Responding to potential risks is further complicated by insufficient information to reliably estimate response costs and by the uncertain availability of federal funding to pay for such actions.

In: Poisoning and Acidification of the Earth's ... ISBN: 978-1-60741-560-2
Editors: Geoffrey E. Mason

Chapter 1

FEELY TESTIMONY ON OCEAN ACIDIFICATION

Richard A. Feely

INTRODUCTION

Good morning Chairman Cantwell and members of the Subcommittee. Thank you for giving me the opportunity to speak with you today on the short- and long-term impacts of ocean acidification on marine resources. My name is Richard Feely, I am a Supervisory Chemical Oceanographer at the Pacific Marine Environmental Laboratory of the National Oceanic and Atmospheric Administration (NOAA) in Seattle, WA. My personal area of research is the study of the oceanic carbon cycle and its impact on marine organisms. I have worked for NOAA for more than 32 years and have published more than 160 peer-reviewed scientific journal articles, book chapters and technical reports. I serve on the U.S. Carbon Cycle Science Program Scientific Steering Group and I am the co-chair of the U.S. Repeat Hydrography Program Scientific Oversight Committee. For today, you have asked me to provide my insights on ocean acidification and its effect on living marine ecosystems. Most of my comments below are derived from the Royal Society Report, *"Ocean Acidification Due to Increasing Atmospheric Carbon Dioxide"* (Raven *et al.*, 2005) and the recent U.S. report, derived from a workshop held jointly by the

National Science Foundation (NSF), NOAA, and the U.S. Geological Survey, entitled *"Impacts of Ocean Acidification on Coral Reefs and Other Marine Calcifiers"* (Kleypas *et al.*, 2006).

OCEAN ACIDIFICATION

Over the past 200 years the release of carbon dioxide (CO_2) from our collective industrial and agricultural activities has resulted in atmospheric CO_2 concentrations that have increased by about 100 parts per million (ppm). The atmospheric concentration of CO_2 is now higher than experienced on Earth for at least the last 800,000 years, and is expected to continue to rise, leading to significant temperature increases in the atmosphere and oceans by the end of this century. The oceans have absorbed approximately 525 billion tons of carbon dioxide from the atmosphere, or about one third of the anthropogenic carbon emissions released during this period (Sabine and Feely, 2007). This natural process of absorption has benefited humankind by significantly reducing the greenhouse gas levels in the atmosphere and minimizing some of the impacts of global warming. However, the ocean's daily uptake of 22 million tons of carbon dioxide is starting to have a significant impact on the chemistry and biology of the oceans. For more than 25 years, NOAA and NSF have co-sponsored repeat hydrographic and chemical surveys of the world oceans, documenting the ocean's response to increasing amounts of carbon dioxide being emitted to the atmosphere by human activities. These surveys have confirmed that the oceans are absorbing increasing amounts of carbon dioxide. Both the hydrographic surveys and modeling studies reveal that the chemical changes in seawater resulting from the absorption of carbon dioxide are lowering seawater pH (Feely *et al.*, 2004; Orr *et al.*, 2005; Caldeira and Wickett, 2005; Feely *et al.*, in press). It is now well established that the pH of our ocean surface waters has already fallen by about 0.1 units from an average of about 8.21 to 8.10 since the beginning of the industrial revolution (on the logarithmic pH scale, 7.0 is neutral (e.g., water), with points higher on the scale being "basic" and points lower being "acidic."). Estimates of future atmospheric and oceanic carbon dioxide concentrations, based on the Intergovernmental Panel on Climate Change (IPCC) CO_2 emission scenarios and general circulation models, indicate that by the middle of this century atmospheric carbon dioxide levels could reach more than 500 parts per million (ppm), and near the end of the century they could be over 800 ppm. This would result in a surface water pH decrease of approximately 0.4 pH units as the ocean becomes more acidic, and the carbonate ion concentration would decrease almost 50 percent by the end of the century (Orr *et al.*, 2005). To put this in historical perspective, this surface ocean pH decrease would result in a pH that is lower than it has been for more than 20 million years (Feely *et al.*, 2004). When CO_2 reacts with seawater, fundamental chemical changes occur that cause a reduction in seawater pH. The interaction between CO_2 and seawater also reduces the availability of carbonate ions, which play an important role in shell formation for a number of marine organisms such as corals, marine plankton, and shellfish. This phenomenon, which is commonly called "ocean acidification," could affect some of the most fundamental biological and geochemical processes of the sea in coming decades. This rapidly emerging issue has created serious concerns across the scientific and fisheries resource management communities.

Effects of Ocean Acidification on Coral Reefs

Many marine organisms that produce calcium carbonate shells studied thus far have shown detrimental effects due to increasing carbon dioxide levels in seawater and the resulting decline in pH. For example, increasing ocean acidification has been shown to significantly reduce the ability of reef-building corals to produce their skeletons, affecting growth of individual corals and making the reef more vulnerable to erosion (Kleypas *et al.*, 2006). Some estimates indicate that, by the end of this century, coral reefs may erode faster than they can be rebuilt. This could compromise the long-term viability of these ecosystems and perhaps impact the thousands of species that depend on the reef habitat. Decreased calcification may also compromise the fitness or success of these organisms and could shift the competitive advantage towards organisms that are not dependent on calcium carbonate. Carbonate structures are likely to be weaker and more susceptible to dissolution and erosion. In long-term experiments corals that have been grown under lower pH conditions for periods longer than one year have not shown any ability to adapt their calcification rates to the low pH levels. In fact, a recent study showed that the projected increase in CO_2 is sufficient to dissolve the calcium carbonate skeletons of some coral species (Fine and Tchernov, 2007).

Effects of Ocean Acidification on Fish and Shellfish

Ongoing research is showing that decreasing pH may also have deleterious effects on commercially important fish and shellfish larvae. Both king crab and silver seabream larvae exhibit very high mortality rates in CO_2-enriched waters (Litzow *et al.*, submitted; Ishimatsu *et al.*, 2004). Some of the experiments indicated that other physiological stresses were also apparent. Exposure of fish to lower pH levels can cause decreased respiration rates, changes in blood chemistry, and changes in enzymatic activity. The calcification rates of the edible mussel (*Mytilus edulis*) and Pacific oyster (*Crassostrea gigas*) decline linearly with increasing CO_2 levels (Gazeau *et al.*, in press). Squid are especially sensitive to ocean acidification because it directly impacts their blood oxygen transport and respiration (Pörtner *et al.*, 2005). Sea urchins raised in lower-pH waters show evidence for inhibited growth due to their inability to maintain internal acid-base balance (Kurihara and Shirayama., 2004). Scientists have also seen a reduced ability of marine algae and free-floating plants and animals to produce protective carbonate shells (Feely *et al.*, 2004; Orr *et al.*, 2005). These organisms are important food sources for other marine species. One type of free-swimming mollusk called a pteropod is eaten by organisms ranging in size from tiny krill to whales. In particular, pteropods are a major food source for North Pacific juvenile salmon, and also serve as food for mackerel, pollock, herring, and cod. Other marine calcifiers, such as coccolithophores (microscopic algae), foraminifera (microscopic protozoans), coralline algae (benthic algae), echinoderms (sea urchins and starfish), and mollusks (snails, clams, and squid) also exhibit a general decline in their ability to produce their shells with decreasing pH (Kleypas *et al.*, 2006).

Effects on Marine Ecosystems

Since ocean acidification research is still in its infancy, it is impossible to predict exactly how the individual species responses will cascade throughout the marine food chain and impact the overall structure of marine ecosystems. It is clear, however, from the existing data and from the geologic record that some coral and shellfish species will be reduced in a high-CO_2 ocean. The rapid disappearance of many calcifying species in past extinction events has been attributed, in large part, to ocean acidification events (Zachos *et al.*, 2005). Over the next century, if CO_2 emissions are allowed to increase as predicted by the IPCC CO_2 emissions scenarios, mankind may be responsible for increasing oceanic CO_2 and making the oceans more corrosive to calcifying organisms than anytime since the last major extinction, over 65 million years ago. Thus, the decisions we make about our use of fossil-fuels for energy over the next several decades will probably have a profound influence on makeup of future marine ecosystems for centuries to millennia.

Economic Impacts

The impact of ocean acidification on fisheries and coral reef ecosystems could reverberate through the U.S. and global economy. The U.S. is the third largest seafood consumer in the world with total consumer spending for fish and shellfish around $60 billion per year. Coastal and marine commercial fishing generates upwards of $30 billion per year and employs nearly 70,000 people (NOAA Fisheries Office of Science and Technology; http://www.st.nmfs.gov/st1/fus/fus05/index.html). Nearly half of the U.S. fishery is derived from the coastal waters surrounding Alaska. Increased ocean acidification may directly or indirectly influence the fish stocks because of large-scale changes in the local ecosystem dynamics. It may also cause the dissolution of the newly discovered deepwater corals in the Alaskan Aleutian Island region. Many commercially important fish species in this region depend on this particular habitat for their survival.

Healthy coral reefs are the foundation of many viable fisheries, as well as the source of jobs and businesses related to tourism and recreation. In the Florida Keys, coral reefs attract more than $1.2 billion in tourism annually. In Hawaii, reef-related tourism and fishing generate $360 million per year, and their overall worth has been estimated at close to $10 billion. In addition, coral reefs provide vital protection to coastal areas that are vulnerable to storm surges and tsunamis.

Conclusions

Ocean acidification may be one of the most significant and far-reaching consequences of the buildup of anthropogenic carbon dioxide in the atmosphere. Results from laboratory, field and modeling studies, as well as evidence from the geological record, clearly indicate that marine ecosystems are highly susceptible to the increases in oceanic CO_2 and the corresponding decreases in pH. Corals and other calcifying organisms will be increasingly affected by a decreased capability to produce their shells and skeletons. Other species of fish and shellfish will also be negatively impacted in their physiological responses due to a decrease in pH levels of their cellular fluids. Because of the very clear potential for ocean-wide impacts of ocean acidification at all levels of the marine ecosystem, from the tiniest phytoplankton to zooplankton to fish and shellfish, we can expect to see significant impacts that are of immense importance to mankind. Ocean acidification is an emerging scientific issue and much research is needed before all of the ecosystems responses are well understood. However, to the limit that the scientific community understands this issue right now, the potential for environmental, economic and societal risk is also quite high, hence demanding serious and immediate attention. For these reasons, the national and international scientific communities have recommended a coordinated scientific research program with four major themes; (1) carbon system monitoring; (2) calcification and physiological response studies under laboratory and field conditions; (3) environmental and ecosystem modeling studies; and (4) socio-economic risk assessments. This research will provide resource managers with the basic information they need to develop strategies for protection of critical species, habitats and ecosystems, similar to what has already been developed for coral reef managers with the publication of *the Reef Manager's Guide* by the U.S. Coral Reef Task Force to help local and regional reef managers reduce the impacts of coral bleaching to coral reef ecosystems.

Thank you for giving me this opportunity to address this Subcommittee. I look forward to answering your questions.

SELECTED REFERENCES

Caldeira, K. & Wickett, M. E. (2005). Ocean model predictions of chemistry changes from carbon dioxide emissions to the atmosphere and ocean. *Journal of Geophysical Research (Oceans)*, *110*, C09S04, doi:10.1029/2004JC002671.

Feely, R. A., Sabine, C. L., Lee, K. Berrelson, W. Kleypas, J., Fabry, V. J. & Millero, F. J. (2004), *Impact of anthropogenic CO_2 on the $CaCO_3$ system in the oceans, Science, 305(5682)*: 362-366.

Feely, R. A., Orr, J., Fabry, V. J., Kleypas, J. A., Sabine, C. L. & Landgon, C. (in press): *Present and future changes in seawater chemistry due to ocean acidification.* AGU Monograph on "The Science and Technology of Carbon Sequestration".

Fine, M. & D. Tchernov (2007). *Scleractinian coral species survive and recover from decalcification, Science, (315),* 1811.

Gazeau, F., Quiblier, C., Jeroen M., Jansen, J. M., Jean-Pierre Gattuso, J.-P., Middelburg, J. J. & H. R. C. Heip (in press) *Impact of elevated CO_2 on shellfish calcification,* Geophysical Research Letters.

Ishimatsu, A., Kikkawa, T., Hayashi, M., Lee, K.-S. & Kita, J. (2004). Effects of CO_2 on marine fish: *Larvae and adults, Journal of Oceanography, Vol. 60*, pp. 731 – 741.

Kleypas, J. A., Feely, R. A., Fabry, V. J., Langdon, C. Sabine, C. L. & Robbins, L. L. (2006). *Impacts of ocean acidification on coral reefs and other marine calcifiers: A guide to future research.* Report of a workshop held 18–20 April 2005, St. Petersburg, FL, sponsored by NSF, NOAA, and the U.S. Geological Survey, 88 pp.

Kurihara, K. & Shirayama, Y. (2004). Impacts of increased atmospheric CO_2 on sea urchin early development, Mar. Ecol;. *Prog. Ser., 274,* 161-169.

Marshall, P. & Schuttenberg, H. (2006). *A Reef Manager's Guide to Coral Bleaching,* Great Barrier Ref Marine Park Authority, Townsville, Australia, 139pp.

Michael A., Litzow, M. A., Short, J. W., J. W. , Persselin, S. L. & Lisa A. Hoferkamp3, L. A. and S. A. Payne, (submitted for publication). Calcite undersaturation reduces larval survival in a crustacean, Science.

Orr, J. C., Fabry, V. J., Aumont, O. Bopp, L., Doney, S. C. Feely, R. A., Gnanadesikan, A., Fruber, N., Ishida, A. Joos, F. Key, R. M., Lindsay, K., Maier-Reimer, E., Matear, R., Monfray, P., Mouchet, A., Najjar, R. G., Plattner, G. K., Rodgers, K. B., Sabine, C. L. Sarmiento, J. L., Schlitzer, R., Slater, R. D., Totterdel, I. J., Weirig, M. F., Yamanaka, Y. & Yool, A. (2005). *Anthropogenic ocean acidification over the twenty-first century and its impact on calcifying organisms, Nature, 437*, 681-686.

Pörtner, H. O., Langenbuch, M. & B. Michaelidis (2005). Synergistic effects of temperature extremes, hypoxia, and increases in CO_2 on marine animals: From Earth history to global change, *J. Geophys.* Res. 110, C09S10, doi:10.1029/2004JC002561.

Raven, J., Caldeira, K., Elderfield, H., Hoegh-Guldberg, O., Liss, P., Riebesell, U., Shepherd, J., Turley, C. & Watson, A. (2005). *Acidification due to increasing carbon dioxide*. In Report 12/05. London , T.R.S.o. (ed.) London : The Royal Society, pp. vii + 60.

Zachos, J. C. Röhl, U., Schellenberg, S. A., Sluijs, A., Hodell, D. A., Keely, D. C. Thomas, E. Nicolo, M., Raffi, I., Lourens, L. J., McCarren, H. & Kroon, D. (2005). Rapid acidification of the ocean during the Paleocene-Eocene thermal maximum, *Science, 308*,1611-16.

In: Poisoning and Acidification of the Earth's ... ISBN: 978-1-60741-560-2
Editors: Geoffrey E. Mason

Chapter 2

MARINE DEAD ZONES: UNDERSTANDING THE PROBLEM

Eugene H. Buck
Natural Resources Policy Resources,
Science, and Industry Division

SUMMARY

An adequate level of dissolved oxygen is necessary to support most forms of aquatic life. While very low levels of dissolved oxygen (hypoxia) can be natural, especially in deep ocean basins and fjords, hypoxia in coastal waters is mostly the result of human activities that have modified landscapes or increased nutrients entering these waters. Hypoxic areas are more widespread during the summer, when algal blooms stimulated by spring runoff decompose to diminish oxygen. Such hypoxic areas may drive out or kill animal life, and usually dissipate by winter. In many places where hypoxia has occurred previously, it is now more severe and longer lasting; in others where hypoxia did not exist historically, it now does, and these areas are becoming more prevalent.

The largest hypoxic area affecting the United States is in the northern Gulf of Mexico near the mouth of the Mississippi River, but there are others as well. Most U.S. coastal estuaries and many developed nearshore areas suffer from varying degrees of hypoxia, causing various environmental damages. Research has been conducted to better identify the human activities that affect the intensity and duration of, as well as the area affected by, hypoxic events, and to begin formulating control strategies.

Near the end of the 105th Congress, the Harmful Algal Bloom and Hypoxia Research and Control Act of 1998 was signed into law as Title VI of P.L. 105-383. Provisions of this act authorize appropriations through NOAA for research, monitoring, education, and management activities to prevent, reduce, and control hypoxia. Under this legislation, an integrated Gulf of Mexico hypoxia assessment was completed in the late 1990s. In 2004, Title I of P.L. 108-456, the Harmful Algal Bloom and Hypoxia Amendments Act of 2004, expanded this authority and reauthorized appropriations through FY2008. Legislation has been introduced in the 110th Congress to reauthorize and amend the Harmful Algal Bloom and Hypoxia Research and Control Act.

As knowledge and understanding have increased concerning the possible impacts of hypoxia, congressional interest in monitoringand addressing the problem has grown. The issue of hypoxia is seen as a search for (1) increased scientific knowledge and understanding of the phenomenon, as well as (2) cost-effective actions that might diminish the size of hypoxic areas by changing practices that promote their growth and development. This report presents an overview of the causes of hypoxia, the U.S. areas of most concern, federal legislation, and relevant federal research programs. This report will be updated as circumstances warrant.

Introduction and Background

Hypoxia refers to a depressed concentration of dissolved oxygen in water. While definitions vary somewhat by region, it is generally agreed that hypoxia in a marine environment occurs seasonally when dissolved oxygen levels fall below 2-3 milligrams per liter. Normal dissolved oxygen concentrations in nearshore marine waters range between 5 and 8 milligrams per liter, and many fish species begin having respiratory difficulties at concentrations below 5 milligrams per liter. In extremely low oxygen environments, less tolerant marine animals cannot survive and either leave the area or die.[1] Mortality is especially likely for sedentary species. In addition, spawning areas and other essential habitat can be destroyed by the lack of oxygen. If these conditions persist, a so-called "dead zone" may develop in which little marine life exists.[2]

The recovery of marine ecosystems following a hypoxic event has not been extensively studied.

Decreased concentrations of dissolved oxygen result in part from natural eutrophication when nutrients (e.g., nitrogen and phosphorus) and sunlight stimulate algal growth (e.g., algae, seaweed, and phytoplankton), increasing the amount of organic matter in an aquatic ecosystem over decades and centuries. As organisms die and sink to the bottom, they are consumed (decomposed) by oxygen-dependent bacteria, depleting the water of oxygen. When this eutrophication is extensive and persistent, bottom waters may become hypoxic, or even anoxic (no dissolved oxygen), while surface waters can be completely normal and full of life. Hypoxia is more likely to occur in coastal waters where the water column is stratified (i.e., layered) because of differences in temperature or salinity or both. Marine dead zones become most noticeable when and where natural eutrophication has been accelerated by human-influenced increases in nutrient loads.

Hypoxia often develops as a result of upwelled ocean waters, particularly along the western coast of the Americas. In many instances, cool, nutrient-rich, deep marine waters rise along the coastal margin and support massive algal blooms that lead to hypoxia. In other instances, upwelled deep water is simply devoid of oxygen. Eutrophication as a result of human activities usually results from non-point sources of nutrients (e.g., runoff from lawns and various agricultural activities including fertilizer use and livestock feedlots), point-source discharge from sewage plants, and emissions from vehicles, power plants, and other industrial sources.

Hypoxic zones frequently occur in coastal areas where rivers enter the ocean (e.g., estuaries). Rivers deliver fresh waters that are rich in nutrients to the saltier estuaries and coastal oceans. The fresh water is less dense than the salt water and typically flows across the top of the sea water. The fresh surface water effectively caps the more dense, saline bottom waters, retarding mixing, creating a two-layer system, and promoting hypoxia development in the lower, more saline waters. In the northern Gulf of Mexico, the greatest algal growth in surface waters occurs about a month after maximum river discharge, with hypoxic bottom water developing a month later.[3] Hypoxia is more likely to occur in estuaries with high nutrient loading and low flushing (i.e., low freshwater turnover).[4] Human activities that increase nutrient loading can increase the intensity, spatial extent, and duration of hypoxic events. Storms and tides may mix the hypoxic bottom water and the aerated surface water, dissipating the hypoxia.

Although the extent of effects of hypoxic events on U.S. coastal ecosystems is still uncertain, the phenomenon is of increasing concern in coastal areas. Several federal agencies are involved in analyzing the problem, including the U.S. Geological Survey (USGS), the National Oceanic and Atmospheric Administration (NOAA), and the U.S. Environmental Protection Agency (EPA). Legislation was enacted by the 105th Congress to provide additional authority and funding for research and monitoring to address these concerns. This authority was extended by the 108th Congress.

Hypoxic Areas

Hypoxic episodes have been recorded in all parts of the world, notably in partially enclosed seas and basins where vertical mixing is minimal, such as the Gulf of Mexico, Chesapeake Bay, the New York Bight, the Baltic Sea, and the Adriatic Sea. In March 2004, the U.N. Environment Program's (UNEP's) *Global Environment Outlook (GEO) Year Book 2003* reported 146 dead zones where marine life could not be supported due to depleted oxygen levels.[5] In 2006, UNEP's Global Programme of Action for the Protection of the Marine Environment from Land-based Activities reported that the frequency and intensity of coastal dead zones is rapidly increasing and could reach 200 sites when a full list of newly identified sites is released in early 2007.[6] Hypoxia has become more frequent and widespread in shallow coastal and estuarine areas.[7] In addition, permanently hypoxic water masses (i.e., oxygen minimum zones) occur in the open ocean, affecting large seafloor surface areas along the continental margins of the eastern Pacific, Indian, and western Atlantic Oceans.[8]

About 21% to 43% of the area of the United States' estuaries have experienced a hypoxic event; more than half of the affected area is in the Mississippi/Atchafalaya River plume.[9] In the Mid-Atlantic region, 13 of 22 estuaries have experienced hypoxic/anoxic events.[10] Of these, the Long Island Sound, Chesapeake Bay, Choptank River, and the New York Bight experience the most serious annual episodes. In the South Atlantic region, hypoxic/anoxic episodes are generally brief,[11] with nearly two-thirds of this region's 21 estuaries experiencing some hypoxia/ anoxia.[12] The Gulf of Mexico region experiences the highest rate of hypoxic/anoxic events, with almost 85% of this region's 38 estuaries experiencing episodes of hypoxia (including the Mississippi/Atchafalaya River plume).[13] The North Atlantic region is not as prone to hypoxic/anoxic events due to the generally low nutrient input (the result of lower population density) and high tidal flushing. However, areas adjacent to high population density (e.g., Cape Cod Bay and Massachusetts Bay) do experience oxygen depletion. In the Pacific region, hypoxia also occurs near population centers (e.g., San Diego Bay, Newport Bay, Alamitos Bay) or in areas of limited circulation, even where water temperatures are cold (e.g., Hood Canal, Whidbey Basin/Skagit Bay).[14]

Gulf of Mexico Dead Zone

The hypoxic zone in the northern Gulf of Mexico is the largest observed in the estuarine and coastal regions of the western hemisphere.[15] First recognized in the early 1970s, it is the largest and most hypoxic area in the United States. The area of hypoxia extends westward from the mouth of the Mississippi River to the upper Texas coast.[16] The seasonal shape and extent of the dead zone are mostly a function of the Mississippi/Atchafalaya River plume, the combined outflow from these two major rivers, and the biological processes it influences. The most reliable predictor of the size of the hypoxic zone is the nitrate-nitrogen load in the two months before the mid-summer mapping cruise.[17] This hypoxic zone generally occurs from May to September, but varies from year to year.

In summer 1993, following massive Mississippi River flooding, the dead zone covered more than 18,000 square kilometers (an area as large as the state of New Jersey); it reached its largest size in summer 2002 — 22,000 square kilometers (an area as large as the state of Massachusetts).[18] Variation in size can be substantial between years — after reaching a maximum size of 20,000 square kilometers in 1999, the dead zone covered a much smaller 4,400 square kilometers in 2000. Although initially predicted to be the largest in size since measurements began, the hypoxic zone in 2007 covered about 20,500 square kilometers. However, a second smaller hypoxic zone of about 4,500 square kilometers also developed off Texas in 2007. This second dead zone was located near Galveston south to Matagorda Bay, and formed after runoff from heavy rains in the Brazos River drainage discharged into the Gulf.[19]

Low velocity winds during the summer result in calm seas that maintain the stratified barrier between surface and bottom water layers. Only during weather disturbances, such as frontal passages, tropical storms, and hurricanes, does vertical mixing of these stratified layers occur. Increased winds and frontal storms in autumn vertically mix the water column, dissipating the hypoxia.[20] In the summer of 1998, this dead zone extended from very near shore (about 10-15 feet water depth) to deeper waters than are normally hypoxic (as much as 160 feet deep off the Mississippi River delta).[21]

Nutrient enrichment is the primary cause of eutrophication, of some algal blooms, and of hypoxia, and is believed to be a major factor in areas such as the northern Gulf of Mexico.[22] The Mississippi watershed drains 41% of the land area of the contiguous 48 states, including most of the farmbelt. Studies of the Mississippi and Atchafalaya Rivers indicate that dissolved nitrogen levels have tripled and phosphorus levels have doubled since 1960, fueling algal growth and the resultant dead zone.[23]

Research suggests that fertilizer leaching and runoff from upriver agricultural sources may be the main sources of nutrients. For example, USGS states that 56% of the Mississippi River's nutrient loading results from fertilizer runoff, with an additional 25% of the Mississippi River nitrogen coming from animal manure (municipal and solid wastes account for 6%, atmospheric deposition for 4%, and unknown sources for 9%).[24] Analysis of cores of sediments underlying the hypoxic area reveals historic information on the Mississippi River watershed, indicating that surface water productivity has increased and bottom water oxygen stress has worsened since the early 1900s, with the most dramatic changes occurring since the 1950s — a change strongly correlated with increased use of commercial fertilizers in the watershed.[25] Hypoxia in the northern Gulf of Mexico was not an obvious or widespread phenomenon prior to the early1970s,[26] but became so since then with the advent of heavy use of artificial fertilizers and changed agricultural practices. Several studies show a direct relationship between river-born nutrients, the highrates of phytoplankton production, and subsequent Gulf of Mexico hypoxia.[27] However, questions remain as to how much of the river's nitrogen might come from natural soil mineralization, what effects floods have on nutrient transport, and how much nitrogen may be contributed by coastal land loss, estimated at 25 square miles per year.[28]

Although studies have found that more than 70% of the total nitrogen transported to the Gulf of Mexico by the Mississippi River originates above the confluence of the Ohio and Mississippi Rivers,[29] focusing on nitrogen runoff per unit area identifies other areas where more concentrated nutrient runoff occurs.[30] Although the lower Mississippi basin (which drains parts of Tennessee, Arkansas, Missouri, Mississippi, and Louisiana) is responsible for only 23% of the nitrogen delivered to the Gulf, some scientists believe that nitrogen removal and/or runoff prevention strategies should focus on this area because of its much greater relative nitrogen contribution[31] and likely more economically efficient nitrogen removal. Researchers estimate that the benefits of nutrient controls in this lower basin could be twice as effective as implementing them in upstream basins.[32] Others dispute that approach, believing that nitrogen removal is much more effective in small streams (i.e., headwaters in drainages) than in large rivers. They contend that, while an area with higher yield per area may seem like a suitable place to focus management attention, focus should be directed to upstream areas where the total yield (regardless of the yield per area) is greater.[33] Workshops and conferences have identified strategies for implementing nutrient controls in the lower Mississippi basin.[34]

Many farming interests maintain that evidence has not proven that agricultural practices are the primary contributors tothe development of the Gulf of Mexico dead zone.[35] Some farmers dispute that they contribute substantially to creating a dead zone that is as much as 1,000 miles away. They argue that their goal is to keep as much as possible of the applied nutrients on their land, since any nutrients that wash away represent wasted money. On the other hand, it is estimated that as much as half of the applied nutrients are lost to surface or ground water and to the air, resulting in approximately $750 million in excess nitrogen (calculated as fertilizer cost) entering the Mississippi River each year.[36]

Impacts of the Gulf of Mexico Dead Zone on Fishing

The Gulf of Mexico supports important, easily accessible commercial and recreational fisheries, bringing in almost $2.9 billion annually in retail sales to Louisiana and supporting almost 50,000 jobs.[37] These highly productive fisheries are the direct result of the input of nutrients from the Mississippi River watershed. Several studies have linked fishery effects, including declines in shrimp yields, with hypoxic episodes and areas in the Gulf. Evidence suggests that the dead zone forces fish and shrimp further offshore as well as into shallow nearshore areas, and reduces the area of essential habitat.[38] Hypoxia increases stress on aquatic ecosystems and may decrease biological diversity in areas experiencing repeated and severe hypoxia.[39] Crowding of marine life into restricted habitat also may lead to indirect consequences through altered competition and predation interactions. In addition, hypoxia may delay or impede the offshore migration of older, larger shrimp, preventing shrimp trawlers from selectively targeting larger shrimp for harvest.

While it is unclear what specific effects the dead zone has on Gulf fisheries, the occurrence of this dead zone may force fishing vessels to changetheir normal fishing patterns, possibly expending more time and fuel to harvest their catch. One study has concluded that any increase in fishing expenses could drive marginal operators out of business. Other potential impacts on Louisiana fisheries include concentration of fishing effort in other areas, resulting in localized overfishing; damage to essential habitat, and possible decreased future production; shellfish mortality, if hypoxic conditions impinge on barrier island beaches and coastal bay waters; localized mortality of finfish and shellfish in shoreline areas; and decreased growth due to reduced food resources in the sediments and water column.[40] The National Oceanic and Atmospheric Administration, Center for Sponsored Coastal Ocean Research, initiated a series of research projects in the northern Gulf of Mexico in 2003 to better understand the effects of hypoxia on fishery resources.

Because the relationship between hypoxic dead zones and populations of commercially and recreationally important living resources is the single most significant scientific barrier to informed management of the hypoxia problem nationally, NOAA's Center for Sponsored Coastal Ocean Research convened a symposium and workshop in March 2007.[41] This meeting focused on the effectiveness of existing approaches for evaluating the effects of hypoxia on ecologically, commercially, and recreationally important fish and shellfish populations in three coastal systems noted for seasonally recurring hypoxic zones — Chesapeake Bay, the Gulf of Mexico, and Lake Erie.

Chesapeake Bay

Hypoxic conditions have been recognized in Chesapeake Bay for many years.[42] In 2003, Virginia Institute of Marine Science (VIMS) scientists found a 250-square-mile area of hypoxic water in upper Chesapeake Bay at depths below about 20 feet, from north of Annapolis nearly to the mouth of the Potomac River.[43] The low oxygen levels were attributed to large nitrogen and phosphorus nutrient inputs,[44] likely carried into the bay by runoff from above-average winter snowfall and spring rains. This runoff was able to pick up nutrients that had accumulated in surrounding soils during four consecutive years of dry weather. These nutrients stimulate large summertime algal blooms in the bay, reducing dissolved oxygen in bay waters when these organisms sink to the bottom and decompose. In 2006, the total hypoxic area in the bay was smaller than in previous years, possibly because spring rainfall was close to the lowest on record, resulting in less runoff and reduced nutrient input to the bay.[45]

Oregon Coast

Although permanently hypoxic water masses (oxygen minimum zones) affect large seafloor surface areas along the continental margin of the eastern Pacific Ocean, no dead zone events had been reported in the nearshore waters off the Oregon coast prior to 2002. Unlike the dead zones in estuarine systems that are caused, in large part, by excessive nutrient runoff from land, the Oregon dead zone forms along the open coast. Coastal winds drive ocean currents that upwell nutrient-rich, but oxygen-poor, waters from the deep sea onto the shallow reaches of the continental shelf.

This upwelling of nutrients fuels phytoplankton blooms that eventually sink and decompose to further reduce oxygen levels in the already low-oxygen waters along the seafloor. Hypoxic zones along the Oregon coast form seasonally and can begin in late spring/early summer in response to the onset of upwelling-favorable winds from the north. This hypoxia can persist through the summer and ultimately recedes during the fall when winds again shift direction and promote ocean currents thatflush low-oxygen water off the continental shelf.

In 2006, the Oregon coastal dead zone was significantly larger (more than 3,000 square kilometers), thicker, longer lasting, and lower in oxygen concentration than in previous years, extending along the ocean floor from Cape Perpetua (Florence) in the south to Cascade Head (Lincoln City) in the north, as close to shore as the 50-foot depth.[46] Underwater surveys of the 2006 event revealed complete disappearance of fish from important habitats in addition to near-complete mortalities of benthic invertebrates that are important in coastal food-webs.[47] The appearance and growth of this dead zone is attributed to fundamental, but not well understood, changes in ocean conditions off the Oregon coast.[48] The severity of the 2006 dead zone appears to reflect (1) changes in ocean and atmosphere conditions that include the strongest declines in offshore source-water oxygen content on record, (2) an exceptional shift in coastal wind patterns that has greatly enhanced upwelling currents, and (3) the persistence of low-oxygen water and phytoplankton blooms along the coast. The extent of the developing hypoxic event in 2007 has not been determined.[49] An analogous dead zone along the open coast of Washington State may also be developing.

While controlling upwelling-caused hypoxic zones is unlikely, increasing the ability to predict and understand the severity and consequences of future hypoxic events will be necessary for managing and ameliorating the social and economic effects of ecosystem changes along the Oregon coast and beyond.

POLICY AND MANAGEMENT EFFORTS

Since a temporary, yet severe, hypoxic event could result in significant mortality or injury to marine mammals, fish, and other aquatic species, many have deemed better understanding and consistent monitoring of hypoxic phenomena necessary. NOAA initiated the Nutrient Enhanced Coastal Ocean Productivity (NECOP) program in 1989 to study the effects of nutrient discharges on U.S. coastal waters.

This study found a clear link between nutrient input, enhanced primary production (i.e., algal and plant growth), and hypoxic events in the northern Gulf of Mexico.[50]

Inresponse to a January 1995 petition from the Sierra Club Legal Defense Fund (currently known as Earthjustice Legal Defense Fund) on behalf of 18 environmental, social justice, and fishermen's organizations, the Gulf of Mexico Program[51] held a conference in December 1995 to outline the issue and identify potential actions. Following that conference, Robert Perciasepe, Assistant EPA Administrator for Water, convened an interagency group of senior Administration officials (the "principals group") to discuss potential policy actions and related science needs. Subsequently, this "principals group" created a Mississippi River/Gulf of Mexico Watershed Nutrient Task Force. Additionally, the White House Office of Science and Technology Policy's Committee on Environment and Natural Resources (CENR) conducted a Hypoxia Science Assessment at the request of EPA. The CENR assessment was peer-reviewed, made available for public comment, and submitted to the task force to assist in developing policy recommendations and a strategy for addressing hypoxia in the northern Gulf of Mexico.

In response to an integrated scientific assessment of hypoxia in the northern Gulf of Mexico by the multi-agency Watershed Nutrient Task Force,[52] a Plan of Action for addressing hypoxia was released in January 2001.[53] This plan's environmental objective is to reduce the five-year running average of the dead zone's area to 5,000 square kilometers or less by the year 2015. Estimates based on water-quality measurements and streamflow records indicate that a 40% reduction in total nitrogen flux to the Gulf is necessary to return to average loads comparable to those during 1955-1970.[54] Model simulations suggest that, short of this 40% reduction, nutrient load reductions of about 20%-30% would result in a 15%-50% increase in dissolved oxygen concentrations in bottom waters. Strategies selected focus on encouraging voluntary, practical, and cost-effective actions; using existing programs, including existing state and federal regulatory mechanisms; and following adaptive management. A reassessment of progress on implementing this action plan was initiated in 2005.[55]

A key consideration is the level and duration of the necessary reduction in excess nutrients from watersheds. Many agricultural lands have been saturated with nutrients for many years, and it may take a long time to "cycle out" excess nitrogen and phosphorus, even if application rates are reduced.[56] While some believe this problem may have no fast solutions and any management regime considered will need to recognize that progress or improvement may not be apparent for years or even decades, others suggest that improved agricultural practices in efficient application of chemical fertilizers and prevention of soil erosion could yield immediate and measurable benefits. One model suggests that a 12% to 14% reduction in the use of fertilizer on croplands in the Mississippi River basin would reduce the net anthropogenic nitrogen inputs by about 30%, without any loss in crop production. Various policy options for modifying agriculture practices continue to be discussed.[57]

Because nonpoint sources are major contributors to the problem at the mouth of the Mississippi River system, many believe the Clean Water Act is the appropriate legal framework for addressing future nutrient inputs. Under §319 of the Clean Water Act, Louisiana and other states have initiated nonpoint-source[58] control programs. These programsseek tocombine local, state, and federal agency resources to address pollution from nonpoint sources within each state.[59] To effectively address concerns, however, nonpoint-source programs would need to be encouraged, funded, and implemented throughout the Mississippi River watershed. Under §303 of the Clean Water Act, states must identify water-quality-limited segments of their waters that are not meeting standards, and then establish total maximum daily loads (TMDLs) for each listed water and each pollutant (e.g., nutrients) that is not meeting current water quality standards. In addition, agricultural research and educational outreach/assistance to farmers might complement regulatory efforts.

Congress took note of the hypoxia problem in 1997 when the conference report on FY1998 Department of the Interior appropriations (H.Rept. 105-337) directed the USGS to give priority attention to hypoxia in its FY1999 budget. Near the end of the 105th Congress, provisions of the Harmful Algal Bloom and Hypoxia Research and Control Act of 1998 were incorporated into the Coast Guard Authorization Act of 1998. This measure was signed into law as P.L. 105-383 on November 13, 1998; Title VI authorized appropriations through NOAA to conduct research, monitoring, education, and management activities for the prevention, reduction, and control of hypoxia, harmful algal blooms, *Pfiesteria*, and other aquatic toxins. In 2004, Title I of P.L. 108-456, the Harmful Algal Bloom and Hypoxia Amendments Act of 2004, expanded this authority and reauthorized appropriations through FY2008. Legislation has been introduced in the 110th Congress to reauthorize and amend the Harmful Algal Bloom and Hypoxia Research and Control Act.[60]

Appropriations and Funding

National Ocean Service

Hypoxia research is regularly funded through appropriations to the National Ocean Service — part of the National Oceanic and AtmosphericAdministration in the Department of Commerce — in their Extramural Research account under *National Centers for Coastal Ocean Science*.[61] For FY2008, NOAA has requested $8.9 million for extramural research grants related to harmful algal bloom and hypoxia, including forecasting.

Department of Agriculture

In the last few years, the U.S. Department of Agriculture's Cooperative State Research, Education, and Extension Service has provided a special research grant of around $220,000 annually to Iowa State University's Leopold Center for Sustainable Agriculture for a project to define and implement new methods and practices in farming that reduce impacts on water quality and the hypoxia problem in the Gulf of Mexico.

Environmental Protection Agency

In the last few years, the Environmental Protection Agency's Environmental Programs and Management account has provided a specific authorization of around $125,000 annually for the Missouri Department of Agriculture's Hypoxia Education and Stewardship Project. This effort seeks to educate Missouri producers about hypoxia and encourage use of practices that will reduce the amount of nitrogen lost through leaching and/or evaporation. The EPA has also funded research on Gulf of Mexico hypoxia through its Gulf Breeze Laboratory

End Notes

[1] Nancy N. Rabalais and R. E. Turner, eds., *Coastal Hypoxia: Consequences for Living Resources and Ecosystems*, Coastal and Estuarine Studies 58 (Washington, DC: American Geophysical Union, 2001), 454 pp.; Lisa A. Levin, "Oxygen Minimum Zone Benthos: Adaptation and Community Response to Hypoxia," *Oceanography and Marine Biology: An Annual Review*, v. 41 (2003): 1-45.

[2] Some molluscs and annelid worms are more tolerant of low oxygen conditions and can survive hypoxic episodes that last many weeks.

[3] D. Justic et al., "Seasonal Coupling Between Riverborne Nutrients, Net Productivity and Hypoxia," *Marine Pollution Bulletin*, v. 26 (1993): 184-189.

[4] R. Turner and N. Rabalais, "Suspended Particulate and Dissolved Nutrient Loadings to Gulf of Mexico Estuaries," in *Biogeochemistry of Gulf of Mexico Estuaries*, T. Bianchi, J. Pennock, and R. Twilley, eds. (New York, NY: John Wiley & Sons, 1999), pp. 89-107.

[5] R. J. Diaz, J. Nestlerode, and M. L. Diaz, "A Global Perspective on the Effects of Eutrophication and Hypoxia on Aquatic Biota" in G. L. Rupp and M. D. White, eds., *Proceedings of the 7th International Symposium on Fish Physiology, Toxicology, and Water Quality*, Tallinn, Estonia, May 12-15, 2003, EPA 600/R-04/049 (Athens, GA: U.S. Environmental Protection Agency, Ecosystems Research Division, 2004); see Box 4 at [http://www.unep.org/geo/yearbook/yb2003/089.htm].

[6] See UNEP press release, available at [http://www.unep.org/Documents.Multilingual/Default.asp?DocumentID=486&ArticleID=5393&l=en].

[7] R. J. Diaz and R. Rosenberg, "Marine Benthic Hypoxia: A Review of Its Ecological Effects and the Behavioral Responses of Benthic Macrofauna," *Oceanography and Marine Biology: An Annual Review*, v. 33 (1995): 245-303. (Hereafter referred to as "Marine Benthic Hypoxia.")

[8] John J. Helly and Lisa A. Levin, "Global Distribution of Naturally Occurring Marine Hypoxia on Continental Margins," *Deep-Sea Research, Part I*, v. 51 (2004): 1159-1168.

[9] N. Rabalais, "Oxygen Depletion in Coastal Waters." *NOAA's State of the Coast Report.* (Silver Spring, MD: NOAA: 1998), National Picture, p. 4, [http://oceanservice.noaa.gov/websites/retiredsites/sotc_pdf/HYP.PDF]. (Hereafter referred to as "Oxygen Depletion in Coastal Waters.")

[10] S. Bricker, "NOAA's National Estuarine Eutrophication Survey: Selected Results for the Mid-Atlantic, South Atlantic and Gulf of Mexico Regions," *Estuarine Research Federation Newsletter*, v. 23, no. 1 (1997): 20-21. (Hereafter referred to as "NOAA's Estuarine Eutrophication Survey: Selected Results.");NOAA's Estuarine Eutrophication Survey, vol. 1: South Atlantic Region (Silver Spring, MD: National Ocean Service, Office of Ocean Resources Conservation and Assessment, 1996), p. 50.

[11] "Oxygen Depletion in Coastal Waters," Regional Contrasts, p. 2.

[12] "NOAA's Estuarine Eutrophication Survey: Selected Results," pp. 20-21.

[13] Ibid., pp. 20-21.

[14] "Oxygen Depletion in Coastal Waters," National Picture, p. 5.

[15] Nancy N. Rabalais, R. E. Turner, and D. Scavia, "Beyond Science into Policy: Gulf of Mexico Hypoxia and the Mississippi River," *BioScience*, v. 52 (2002): 129-142.

[16] White House Office of Science and Technology Policy, Committee on Environment and Natural Resources, Hypoxia Work Group, *Gulf of Mexico Hypoxia Assessment Plan* (March 1998), p. 3. (Hereafter referred to as Hypoxia Work Group.)

[17] R. E. Turner, Nancy N. Rabalais, and D. Justic, "Predicting Summer Hypoxia in the Northern Gulf of Mexico: Riverine N, P and Si Loading," *Marine Pollution Bulletin*, v. 52 (2006): 139-148.

[18] Nancy N. Rabalais, R. E. Turner, and W. J. Wiseman, Jr., "Hypoxia in the Gulf of Mexico, a.k.a. 'The Dead Zone,'" *Annual Review of Ecology and Systematics*, v. 33 (2002): 235-263. (Hereafter referred to as "Hypoxia, a.k.a. 'The Dead Zone.'")

[19] Jeff Franks, "Gulf "Dead Zone" Reaches Record Size," *Reuters*, August 3, 2007 [http://www.ocregister.com/ocregister/sciencetech/homepage/article_1797591.php], and Eric Berger, "Texas 'Dead Zone' is Dissipating, *Houston Chronicle*, August 10, 2007 [http://www.chron.com/disp/story.mpl/metropolitan/5045528.html].

[20] In August-September 2005, the winds from Hurricanes Katrina and Rita likely promoted a somewhat earlier dissipation of the dead zone off the mouth of the Mississippi River, although Hurricane Cindy in early July 2005 resulted in a smaller mid-summer hypoxic zone than expected that summer.

[21] "Hypoxia, a.k.a. 'The Dead Zone.'"

[22] Hypoxia Work Group, p. 2.

[23] R. E. Turner and N. Rabalais, "Changes in Mississippi River Quality This Century — Implications for Coastal Food Webs," *BioScience*, v. 41, no. 3 (1991): 140-147; R. E. Turner and Nancy N. Rabalais, "Suspended Sediment, C, N, P, and Si Yields from the Mississippi River Basin," *Hydrobiologia*, v. 511 (2004): 79-89; R. E. Turner, et al., "Future Aquatic Nutrient Limitations," *Marine Pollution Bulletin*, v. 46 (2003): 1032-1034.

[24] R. H. Meade, ed., *Contaminants in the Mississippi River*, 1987-92, Circular 1133 (Denver, CO: U.S. Geological Survey, 1995), 140 pp.; D. A. Goolsby et al., F*lux and Sources of Nutrients in the Mississippi-Atchafalaya River Basin*, Topic 3 Report for the Integrated Assessment of Hypoxia in the Gulf of Mexico, NOAA Coastal Ocean Program Decision Analysis Series No. 17 (Silver Spring, MD: NOAA Coastal Ocean Program, 1999).

[25] T. A. Nelson et al., "Time-Based Correlation of Biogenic, Lithogenic and Authigenic Sediment Components with Anthropogenic Inputs in the Gulf of Mexico NECOP Study Area," *Estuaries*, v. 17 (December 1994): 873; B. J. Eadie et al., "Records of Nutrient-Enhanced Coastal Productivity in Sediments from the Louisiana Continental Shelf."*Estuaries*, v. 17 (December 1994): 754-765; N. Rabalais et al., "Nutrient Changes in the Mississippi River and System Responses on the Adjacent Continental Shelf," *Estuaries*, v. 19 (1996): 386-407; S. Gupta et al., "Seasonal Oxygen Depletion in Continental-Shelf Waters of Louisiana: Historical Record of Benthic Foraminifers," *Geology*, v. 24 (1996): 227-230; and R. E. Turner and N. Rabalais, "Coastal Eutrophication Near the Mississippi River Delta," *Nature*, v. 368 (1994): 619-621.

[26] N. Chen et al., "Historical Trends of Hypoxia on the Louisiana Shelf: Applications of Pigments as Biomarkers," *Organic Geochemistry*, v. 32 (2001): 543-561; D. Justic, N. N. Rabalais, and R. E. Turner, Modeling the Impacts of Decadal Changes in Riverine Nutrient Fluxes on Coastal Eutrophication Near the Mississippi River Delta," *Ecological Modeling*, v. 152 (2002): 33-46; L. E. Osterman, "Benthic Foraminifers from the Continental Shelf and Slope of the Gulf of Mexico: An Indicator of Shelf Hypoxia," Estuarine, *Coastal and Shelf Science*, v. 58 (2003): 17-35; and N. N. Rabalais et al., "Ecosystem History of Mississippi River-Influenced Continental Shelf Revealed Through Preserved Phytoplankton Pigments,: Marine Pollution Bulletin, v. 49 (2004): 537-547.

[27] F. H. Sklar and R. E. Turner, "Characteristics of Phytoplankton Production Off Barataria Bay in an Area Influenced by the Mississippi River," *Marine Science*, v. 24 (1981): 93-106; S.

E. Lohrenz, M. J. Dagg, and T. E. Whitledge, "Enhanced Primary Production at the Plume/Oceanic Interface of the Mississippi River," *Continental Shelf Research*, v. 7 (1990): 639-664; S. E. Lohrenz et al., "Variations in Primary Productivity of Northern Gulf of Mexico Continental Shelf Waters Linked to Nutrient Inputs from the Mississippi River," *Marine Ecology Progress Series*, v. 155 (1997): 435-454; D. Justic, Nancy N. Rabalais, and R. E. Turner, "Effects of Climate Change on Hypoxia in Coastal Waters: A Doubled CO_2 Scenario fo the Northern Gulf of Mexico," *Limnology and Oceanography*, v. 41, no. 5 (1996): 992-1003.

28 D. Malakoff, "Death by Suffocation in the Gulf of Mexico," Science, v. 281 (July 10, 1998): 190-192. (Hereafter referred to as "Death by Suffocation.")

29 R. Alexander, R. Smith, and G. Schwarz, "The Regional Transport of Point and Nonpoint-Source Nitrogen to the Gulf of Mexico," *Proceedings of the First Gulf of Mexico Hypoxia Management Conference*, Kenner, LA (December 5-6, 1995), pp. 127-133 (hereafter referred to as "Regional Transport"); R. H. Meade, ed., *Contaminants in the Mississippi River*, 1987-92, Circular 1133 (Denver, CO: U.S. Geological Survey, 1995), 140 pp.; D. A. Goolsby et al., *Flux and Sources of Nutrients in the Mississippi-Atchafalaya River Basin*, Topic 3 Report for the Integrated Assessment of Hypoxia in the Gulf of Mexico, NOAA Coastal Ocean Program Decision Analysis Series No. 17 (Silver Spring, MD: NOAA Coastal Ocean Program, 1999).

30 Of this total, 39% is contributed by the upper and central Mississippi River basins (which include Minnesota, Wisconsin, Iowa, Missouri, and Illinois), 22% by the Ohio River basin, and 11% by the Missouri River basin.

31 Nitrogen runoff for the lower Mississippi basin is 2,072 kilograms of nitrogen per square kilometer per year compared to 708 kilograms of nitrogen per square kilometer per year for the upper Mississippi basin and 437 kilograms of nitrogen per square kilometer per year for the Ohio River basin.

32 "Regional Transport," p. 131.

33 R. B. Alexander, R. A. Smith, and G. E. Swartz, "Effect of Stream Channel Size on the Delivery of Nitrogen to the Gulf of Mexico," *Nature*, v. 403 (2000): 758-761; R. B. Alexander and R. A. Smith, "Trends in the Nutrient Enrichment of U.S. Rivers During the Late 20th Century and Their Relation to Changes in Probable Stream Trophic Conditions," *Limnology and Oceanography*, v. 51 (2006): 639-654.

34 For example, see C. L. Cordes and B. A. Vairin, eds., *Workshop on Solutions and Approaches for Alleviating Hypoxia in the Gulf of Mexico*, NWRC Special Report 98-02 (Lafayette, LA: U.S. Geological Survey, 1998), 53 p.

35 C. David Kelly, "Hypoxia Issue Paints a Murky Picture," *Voice of Agriculture, American Farm Bureau* (September 29, 1997), at [http://www.fb.org/index.php?fuseaction=newsroom.focusfocus&year=1997&file=fo0929.html].

36 "Death by Suffocation," pp. 190-192.

37 Southwick Associates, *The Economic Benefits of Fisheries, Wildlife and Boating Resources in the State of Louisiana* (Arlington, VA: March 1997), 21 pp.

38 J. A. Downing et al., *Gulf of Mexico Hypoxia: Land-Sea Interactions*, Council for Agricultural Science and Technology, Task Force Report No. 134 (1999), 40 p.; N. N. Rabalais and R. E. Turner, eds., *Coastal Hypoxia: Consequences for Living Resources and Ecosystems*, Coastal and Estuarine Studies 58 (Washington, DC: American Geophysical Union, 2001), 454 p.; J. K. Craig, L. B. Crowder, and T. A. Henwood, "Spatial Distribution of Brown Shrimp (*Farfantepenaeus aztecus*) on the Northwestern Gulf of Mexico Shelf: Effects of Abundance and Hypoxia," *Canadian Journal of Fisheries and Aquatic Science*, v. 62 (2005): 1295-1308; J. K. Craig and L. B. Crowder, "Hypoxia-Induced Habitat Shifts and Energetic Consequences in Atlantic Croaker and Brown Shrimp on the Gulf of Mexico Shelf," *Marine Ecology Progress Series*, v. 294 (2005): 79-94.

[39] "Marine Benthic Hypoxia," pp. 285-287; N. N. Rabalais and R. E. Turner, eds., *Coastal Hypoxia: Consequences for Living Resources and Ecosystems*, Coastal and Estuarine Studies 58 (Washington, DC: American Geophysical Union, 2001), 454 pp.

[40] J. Hanifen et al., "Potential Impacts of Hypoxia on Fisheries: Louisiana's Fishery-Independent Data," *Proceedings for the First Gulf of Mexico Hypoxia Management Conference*, Kenner, LA (December 5-6, 1995), pp. 87-100; N. N. Rabalais and R. E. Turner, eds., *Coastal Hypoxia: Consequences for Living Resources and Ecosystems*, Coastal and Estuarine Studies 58 (Washington, DC: American Geophysical Union, 2001), 454 pp.

[41] For more information on this meeting, see [http://www.cop.noaa.gov/stressors/pollution/features/fs-2007-03-16-hyp.html].

[42] Denise L. Breitburg, "Episodic Hypoxia in Chesapeake Bay: Interacting Effects of Recruitment, Behavior, and Physical Disturbance," *Ecological Monographs*, v. 62, no. 4 (1992): 525-546; James D. Hagy et al., "Hypoxia in Chesapeake Bay, 1950-2001: Long-Term Change in Relation to Nutrient Loading and River Flow," *Estuaries*, v. 27, no. 4 (2004): 634-659.

[43] See the VIMS press release at [http://www.vims.edu/newsmedia/press_release/hypoxia. html].

[44] For more information on nutrient input to Chesapeake Bay, see [http://chesapeake.usgs. gov/features.html].

[45] See "State of the Bay 2006" at [http://wrc.iewatershed.com/chesapeake/SOTB_2006.pdf].

[46] For more information, see [http://www.piscoweb.org/research/oceanography/hypoxia] and[http://oregonstate.edu/dept/ncs/newsarch/2006/Oct06/deadzoneends.html]

[47] See [http://oregonstate.edu/events/newsevents/deadzone_video.html].

[48]Brian A. Grantham, et al., "Upwelling-Driven Nearshore Hypoxia Signals Ecosystem andOceanographic Changes in the Northeast Pacific," *Nature*, v. 429 (June 17, 2004): 749-754.

[49] See [http://www.cbbulletin.com/Free/231315.aspx].

[50] NOAA, Coastal Ocean Program Office, *Nutrient-Enhanced Coastal Ocean Productivity, Proceedings of 1994 Synthesis Workshop* (1995), p. 119; see also *Estuaries*, v. 17, no. 4 (December 1994): 729-911.

[51] The Gulf of Mexico Program is a cooperative federal-state effort beginning after Congress, through P.L. 102-178, designated 1992 as the Year of the Gulf of Mexico. For additional information on this program, see [http://www.epa.gov/gmpo/].

[52] See [http://www.nos.noaa.gov/products/pubs_hypox.html].

[53] See [http://www.epa.gov/msbasin/taskforce/pdf/actionplan.pdf].

[54] D. Scavia, et al., "Predicting the Response of Gulf of Mexico Hypoxia to Variations in Mississippi River Nitrogen Load," *Limnology and Oceanography*, v. 48 (2003): 951-956.

[55] See [http://www.epa.gov/msbasin/taskforce/reassess2005.htm].

[56] "Death by Suffocation" (citing Don Goolsby, USGS, Denver, CO), pp. 190-192.

[57] For example, see Suzie Greenhalgh and Amanda Sauer, "Awakening the Dead Zone: An Investment for Agriculture, Water Quality, and Climate Change," *World Resources Institute Issue Brief* (February 2003), 24 p.

[58] Originating from land use activities; sediment, organic and inorganic chemicals, and biological, radiological, and other toxic substances are carried to lakes and streams by surface runoff.

[59] D. Sabin and J. Boydstun, "Louisiana Activities and Programs in Nutrient Control and Management," *Proceedings of the First Gulf of Mexico Hypoxia Management Conference*, Kenner, LA (December 5-6, 1995), pp. 196-198.

[60] For additional information on action in the 110th Congress on this issue, see CRS Report RL33813, *Fishery, Aquaculture, and Marine Mammal Legislation in the 110th Congress*, by Eugene H. Buck.

[61] More detail on NOAA programs and funded studies can be found at [http://www.cop. noaa.gov/stressors/extremeevents/hab/features/hypoxiafs_report1206.html].

In: Poisoning and Acidification of the Earth's ... ISBN: 978-1-60741-560-2
Editors: Geoffrey E. Mason

Chapter 3

OCEAN ACIDIFICATION

Eugene H. Buck[1] and Peter Folger[2]
[1]Natural Resources Policy
[2]Energy and Natural Resources Policy

SUMMARY

With increasing concentrations of carbon dioxide (CO_2) in the atmosphere, the extent of effects on the ocean and marine resources is an increasing concern. One aspect of this issue is the potential for seawater to become increasing acidic (i.e., ocean acidification). Scientists are concerned that increasing acidity could result in reduced growth or even death of shell-forming animals (e.g., molluscs and certain planktonic organisms) as well as disruption of marine food webs and reproductive physiology. Congress is beginning to focus attention on better understanding this issue and determining how it might be addressed.

WHAT IS OCEAN ACIDIFICATION?

As increasing carbon dioxide (CO_2) from the atmosphere dissolves in seawater, seawater chemistry is altered. The prevailing pH of the ocean is around 8.1, or slightly alkaline.[1] Ocean acidification is the name given to the process whereby pH decreases as seawater becomes more acidic (i.e., less alkaline) when increasing amounts of anthropogenic CO_2 from the atmosphere dissolve in seawater to form carbonic acid. Scientists are concerned that increasing acidity could disrupt physiological processes of marine organisms and detrimentally alter marine ecosystem relationships.

AT WHAT RATE IS OCEAN ACIDIFICATION OCCURRING AND WHAT FACTORS AFFECT THIS RATE?

Of the approximately 7 billion metric tons of carbon that all the countries in the world release as CO_2 into the atmosphere each year, the oceans take up about 2 billion tons.[2] Between pre-industrial times and 1994, average surface ocean pH is estimated to have decreased (i.e., acidity increased) almost 0.1 pH unit.[3] That increase sounds small, but it represents an increase of 20% in the concentration of hydrogen ions, because the pH scale is logarithmic (i.e., a pH of 7 is ten times more acidic than a pH of 8, and one hundred times as acidic than a pH of 9). More recent research indicates that the rate of acidification, at least in some locations, may be proceeding much more rapidly.[4] As long as atmospheric CO_2 continues to increase, many scientists expect the oceans to continue to become more acidic.

Gases, such as CO_2, are generally less soluble in warm water than in cold water. Thus, marine waters near the poles have a much greater capacity for dissolving CO_2 than do ocean waters in the tropics. It follows that the rate of ocean acidification, and the appearance of impacts on physical and biological features, initially may be more likely to occur in marine waters nearer the poles.

Key scientific questions concern which factors may affect the future rate of acidification, especially whether the rate of increase will remain constant in direct relationship to the amount of CO_2 in the atmosphere or whether other factors will result in an acceleration or deceleration of this acidification rate. There is also the question of equilibrium—that is, how long will it take the acidification process of ocean waters to come into equilibrium with the concentration of atmospheric CO_2? An adjunct to this question is how long will it take the rate of ocean acidification to slow in response to any measures that might be taken to slow or halt the increasing concentration of CO_2 in the atmosphere?

What Are Some of the Potential Effects of Ocean Acidification?[5]

Even with increasing concentrations of atmospheric CO_2, the oceans are not likely to reach pH values of less than 7 (neutral). But as CO_2 dissolves into the ocean's surface waters, it changes the proportions of ions available to marine organisms who make biogenic calcium carbonate to form shells. As more CO_2 dissolves into the surface ocean, not only are some organisms hindered in their ability to make calcium carbonate, but calcium carbonate skeletons of marine organisms may actually start to dissolve at or near the ocean surface in some parts of the globe.

These changes are occurring because of the complex interplay between rising CO_2 levels in the atmosphere and the ocean's chemistry, a process that scientists have recognized for decades. When atmospheric CO_2 dissolves into the ocean, it forms carbonic acid (H_2CO_3). Some of the carbonic acid dissociates in ocean waters producing hydrogen ions (H^+) and bicarbonate ions (HCO_3^-). As the number of hydrogen ions increases, the pH of the ocean gets lower, or more acidic. When more CO_2 is added to the atmosphere, more carbonic acid forms in the ocean. Over the past several decades, about half of the CO_2 released by human activities has remained in the atmosphere; of the remainder, about 30% has entered the oceans.[6] As a result, the additional carbonic acid has increased average ocean acidity by approximately 0.1 pH unit (i.e., an increase of 20% in hydrogen ion concentration).[7]

A lower pH affects marine life in the oceans and is related to other changes in ocean chemistry. For example, the bicarbonate ion (HCO_3^-) mentioned above further dissociates into H^+ and carbonate ions (CO_3^{2-}), and it is the abundance and availability of carbonate ions that are critical to many shell-forming marine organisms. At current average ocean pH levels (about 8 or above), ocean surface waters have ample carbonate ions to form shells and grow corals. However, as increased amounts of carbonic acid form in the ocean from higher CO_2 levels in the atmosphere, pH gets lower and the amount of carbonate ion in the oceans decreases, resulting in less carbonate ion available for making shells.

Organisms make biogenic calcium carbonate for their shells by combining calcium ions (Ca^{2+})— which are abundant in the oceans—with carbonate ions to form solid calcium carbonate ($CaCO_3$). Marine organisms such as corals and pteropods use one mineral type of calcium carbonate called aragonite, and other organisms such as foraminifera and coccolithophorids use another type called calcite. Under present conditions of ocean chemistry, both forms of calcium carbonate are relatively stable in the surface ocean. The surface ocean is deemed saturated with respect to both calcite and aragonite, meaning that organisms can form shells from either mineral type. However, as more carbonic acid is added to the surface ocean from higher levels of CO_2 in the atmosphere, the level of saturation decreases. If the ocean waters become undersaturated, then shells made from aragonite or calcite would tend to dissolve. Shells made from aragonite would tend to dissolve first, at lower concentrations of carbonic acid (and thus at higher pH values) than would shells made from calcite.

Although surface waters currently are more than fully saturated with carbonate, declining carbonate concentrations (caused by increasing acidity) are projected to reduce the ability of organisms to form biogenic calcium carbonate. Some suggest that marine surface waters closer to the poles may become undersaturated within the next 50 years.[8] Researchers at the Antarctic Climate and Ecosystems Cooperative Research Centre have demonstrated significant reductions in shell mass and thickness of several Southern Ocean marine plants and animals that appear consistent with the projected effects of recent increased acidification of the ocean.[9]

In response to ocean acidification, scientists have projected that mussel and oyster calcification, and thus shell strength, could decrease by 25% and 10%, respectively, by the end of the 21st century, according to the Intergovernmental Panel on Climate Change's IS92a scenario.[10] There is also the concern that increased acidification may cause marine calcium carbonate sediments to dissolve with potential detrimental effects on species and communities residing in and on these sediments.[11] Since many of these organisms provide food or modify habitat for other organisms, the well-being of these carbonate-dependent species will affect other species. Because of these interrelationships, the potential indirect effects of acidification on other marine organisms is not well understood. Others raise the concern that ocean acidification, by negatively affecting calcifying plankton species, could shift ecological balances so as to increase populations of noncalcifying species, such as jellyfish.[12]

There are also concerns that increasing acidification of ocean waters could alter the ability of some organisms to conduct essential biochemical and physiological processes. For example, increased CO_2 concentration appears to cause physiological changes that can be detrimental to the relationship between coral and their symbiotic zooxanthellae,[13] wherein the coral host loses control over the zooxanthellae, resulting in uncontrolled division.[14] Evidence suggests that larval and juvenile fish may be particularly susceptible to changes in ocean pH levels.[15] In addition, scientists have found that, when exposed to water of pH 7.7, roughly equivalent to acidity levels predicted for the year 2100, sea urchin sperm swam much more slowly. Overall, fertilization fell by 25%, and in almost 26% of cases where the eggs were fertilized, they did not survive long enough to develop into larvae.[16]

In nature, the ocean waters at depths of hundreds or thousands of feet become undersaturated with respect to aragonite and calcite, which is why most of the shells from dead organisms that "rain" down from the ocean surface dissolve before reaching the ocean floor.[17] One concern is that increasing levels of atmospheric CO_2 entering the ocean and mixing with deeper waters will result in a shoaling of undersaturated conditions, which could reach shallow levels where most of the shell-forming organisms live. Under those conditions, some plankton and corals may have difficulty maintaining their calcium carbonate skeletons. This is likely to occur first in colder waters near the poles, which tend to have higher levels of dissolved CO_2 than warmer waters near the equator. Recent research suggests that some areas, such as the Southern Ocean, could become undersaturated with respect to aragonite by 2050 or sooner.[18]

The fossil record indicates that marine organisms may be quite sensitive to ocean acidification—about 55 million years ago, a large injection of CO_2 into the deep ocean, presumably resulting from a massive methane release, was followed by the extinction of some species of benthic foraminifera.[19] Others caution that these paleo-events may be imperfect analogs to current conditions.[20]

Some have raised questions downplaying the potential harm to coral reefs from ocean acidification. Differences of opinion exist on the relative effects of climate change as expressed in increased CO_2 when compared to increased ocean temperature. Opinion has been expressed that, in marine systems, increased temperature may have detrimental effects comparable to or larger than those seen from increased CO_2 concentration, for corals and for phytoplankton.[21] Although calcification rates in massive *Porites* coral were reported to have declined over a 16-year study period by approximately 21% in two regions on Australia's Great Barrier Reef, these findings were consistent with studies of the synergistic effect of elevated seawater temperatures and CO_2 concentrations on coral calcification.[22]

As noted above in relation to corals, scientists point out that CO_2 is far less soluble in warm water than in cold water. Thus the cold polar and deeper waters will likely be the first to be affected by ocean acidification,[23] while the tropical surface waters and coral reefs may be among the last places to be threatened by the effects of ocean acidification. Also supportive of the ability of certain corals to survive decreasing pH, stony and soft corals have been grown successfully in open systems with low pH water from a saltwater well since the 1970s.[24]

What Are the Natural and Human Responses That Might Limit or Reduce Ocean Acidification?

Our ability to reduce ocean acidification through artificial means is unproven. Proposals have suggested the addition of chemicals to the ocean, such as (1) using iron compounds to stimulate planktonic algae growth whereby the increased photosynthesis might capture/remove dissolved CO_2, or (2) using limestone to neutralize (i.e., buffer) the more acidic streams and rivers near where they enter oceans and close to sources of limestone. Unless a massive global effort is mounted, these techniques will at best be effective only on a very local scale. In addition, manipulation of ocean chemistry has the potential to damage or otherwise alter the marine environment and ecosystems. Reducing CO_2 emissions to the atmosphere and/or removing CO_2 from the atmosphere (i.e., carbon sequestration) currently appear to be the only practical ways to minimize the risk of large-scale and long-term changes to the acidity of marine waters. Because of the continuing increase in CO_2 levels in the atmosphere, and its residence time there, acidification of the oceans will likely continue for a long time. Even if atmospheric CO_2 were to return to pre-industrial levels, it would likely take tens of thousands of years for ocean chemistry to return to a condition similar to that occurring at pre-industrial times more than 200 years ago.[25]

What Is the Federal Government Doing about Ocean Acidification?

Much of the current federal attention to ocean acidification focuses on research to better understand the chemical processes involved and to become better able to predict how ocean ecosystems might respond to increasing acidification.

A variety of programs conducted within the National Oceanic and Atmospheric Administration (NOAA) help to gain a better understanding of ocean acidification.[26] The Pacific Marine Environmental Laboratory's CO_2 shipboard measurements and monitoring buoys provide data that help discern seasonal changes in the oceanic carbon system. The Atlantic Oceanographic and Meteorological Laboratory monitors changes in CO_2 and pH through the use of chemical sensors on ships and moorings. NOAA's Repeat Hydrography Program provides data on the large-scale changes of carbon system and ocean acidification over decadal time scales. Sea Grant supports research on the effects of ocean acidification on coral reefs in Hawaii. NOAA's Geophysical Fluid Dynamics Laboratory participated in the Ocean-Carbon Cycle Model Intercomparison Project (OCMIP2) to develop an international collaboration to improve the predictive capacity of carbon cycle models. NOAA Fisheries Alaska Fisheries Science Center has been conducting exposure studies of blue king crab larval survival due to reduced pH and has developed an ocean acidification research plan.[27] NOAA Fisheries Southwest Fisheries Science Center has been evaluating the long-term impacts of low pH on marine plankton in the California Current and off Antarctica. Projects funded by NOAA Global Carbon Cycle program at NOAA laboratories and universities provide information to address the CO_2 and pH changes in the ocean. NOAA estimates that its annual expenditures for these ocean acidification research and related programs in FY2008 were about $4.3 million; for FY2009, NOAA has requested $4.06 million.[28]

In October 2008, NOAA and the National Science Foundation commissioned an 18-month comprehensive national study by the National Research Council of the National Academy of Sciences of how CO_2 emissions absorbed into the oceans may be altering fisheries, marine mammals, coral reefs, and other natural resources.[29] This study was required by Section 701 of P.L. 109-479.[30]

The National Aeronautics and Space Administration (NASA) has a number of space projects that contribute to a better understanding of ocean acidification.[31]

WHAT IS THE CONGRESSIONAL INTEREST IN OCEAN ACIDIFICATION?

Congressional attention is focused primarily on addressing the cause of ocean acidification— increasing atmospheric CO_2.[32] Legislative attention to ocean acidification focuses on authorizing and funding research to increase knowledge about ocean acidification and its potential effects on marine ecosystems.

In comparison to previous sessions of Congress, legislative interest in ocean acidification expanded significantly in the 110th Congress. In the 110th Congress, the Senate Commerce, Science, and Transportation Subcommittee on Oceans, Atmosphere, Fisheries, and Coast Guard held a hearing on the effects of climate change and ocean acidification on living marine resources. In the House, the Committee on Science and Technology held a hearing on H.R. 4174, the Federal Ocean Acidification Research and Monitoring Act (establishing an interagency committee to develop an ocean acidification research and monitoring plan and establishing an ocean acidification program within NOAA). The House later passed this measure, but it was not considered in the Senate. In addition, the House passed H.R. 3221 (amended), wherein Section 7471 would have directed the Secretary of Commerce to develop a national strategy to predict, plan for, and mitigate climate change effects, including ocean acidification, on ocean and coastal ecosystems to ensure the recovery, resiliency, and health of these systems. The Senate did not consider this measure. The Senate did pass H.R. 3093, after inserting language that would have specified NOAA funds to initiate the study of ocean acidification and its effects, required by Section 701 of P.L. 109-479; however, this language was deleted in conference on this measure. More than a dozen other bills containing provisions on this issue did not receive any floor action in the 110th Congress.[33]

Author Contact Information

Eugene H. Buck
Specialist in Natural Resources Policy
gbuck@crs.loc.gov, 7-7262

Peter Folger
Specialist in Energy and Natural Resources Policy
pfolger@crs.loc.gov, 7-1517

End Notes

[1] The pH scale is an inverse logarithmic representation of hydrogen proton (H+) concentration, indicating the activity of hydrogen ions (or their equivalent) in the solution. A pH of less than 7.0 is considered acidic, while a pH greater than 7.0 is considered basic (alkaline); a pH level of 7.0 is defined as "neutral."

[2] Dr. Richard A. Feely, Pacific Marine Environmental Laboratory, National Oceanic and Atmospheric Administration, U.S. Department of Commerce, World Ocean Forum, Nov. 13-14, 2006, at http://www.thew2o.net/events/oceans oa_q_and_a.php.

[3] James C. Orr et al., "Anthropogenic Ocean Acidification over the Twenty-First Century and Its Impact on Calcifying Organisms," *Nature*, vol. 437 (2005): 681-686.

[4] J. Timothy Wootton, Catherine A. Pfister, and James D. Forester, "Dynamic Patterns and Ecological Impacts of Declining Ocean pH in a High-Resolution Multi-Year Dataset," *Proceedings of the National Academy of Sciences of the United States of America*, v. 105, no. 48 (Dec. 2, 2008): 18848-18853.

[5] Since the marine environment is complex and some of the likely changes may be years in the future, the potential effects identified in this section, although many are supported by laboratory experimentation, are primarily conjecture and/or forecasts.

[6] Richard A. Feely et al., "Impact of Anthropogenic CO_2 on the CaCO3 System in the Oceans," *Science* (2004), vol. 305, pp. 362-366.

[7] James C. Orr et al., "Anthropogenic Ocean Acidification over the Twenty-First Century and Its Impact on Calcifying Organisms," *Nature*, vol. 437 (2005): 681-686.

[8] Orr et al. (2005).

[9] Bruce Mapstone, "Acid Oceans in the Spotlight," *Antarctic Climate and Ecosystem News*, edition 4 (August 2008): 1.

[10] Frederic Gazeau et al., "Impact of Elevated CO_2 on Shellfish Calcification," *Geophysical Research Letters*, v. 34, no. 7 (Apr. 16, 2007): L07603 (5 p.).

[11] M. Gehlen, L. Bopp, and O. Aumont, "Short-term Dissolution Response of Pelagic Carbonate Sediments to the Invasion of Anthropogenic CO_2: A Model Study," *Geochem. Geophys. Geosyst.*, v. 9 (Feb. 16, 2008): Q02012.

[12] A. J. Richardson and M. J. Gibbons, "Are Jellyfish Increasing in Response to Ocean Acidification?," *Limnology and Oceanography*, v. 53, no. 5 (2008):2040-2045.

[13] Algae that live in the tissues of a variety of invertebrate groups (e.g., corals). Under lighted conditions, zooxanthellae use the CO_2 and waste materials produced by corals, providing oxygen and food substances that may be used by the coral.

[14] J. Ravindran and Chandralata Raghukumar, "Pink-Line Syndrome, a Physiological Crisis in the Scleractinian Coral Porites lutea," *Marine Biology*, v. 149, no. 2 (2006): 347-3 56.

[15] A. Ishimatsu et al., "Effects of CO_2 on Marine Fish: Larvae and Adults," *Journal of Oceanography*, v. 60, no. 4 (2004): 73 1-741.

[16] Jon N. Havenhand, Fenina-Raphaeia Buttler, Michael C. Thorndyke, and Jane E. Williamson, "Near-Future Levels of Ocean Acidification Reduce Fertilization Success in a Sea Urchin," Current Biology, v. 18, no. 15 (August 2008): 651-652.

[17] Recent research suggests that no more than about 30% of the calcium carbonate produced in the surface ocean each year is buried in shallow or deep sea sediments; the rest dissolves on its way down the water column. See Feely et al. (2004), p. 365.

[18] Orr et al. (2005).

[19] J. C. Zachos et al., "Rapid Acidification of the Ocean During the Paleocene-Eocene Thermal Maximum," *Science*, v. 308 (2005): 1611-1615.

[20] Scott C. Doney et al., "Ocean Acidification: The Other CO_2 Problem," *Annual Review of Marine Science*, v. 1 (January 2009): 169-192.

[21] Clinton E. Hare et al., "Consequences of Increased Temperature and CO_2 for Phytoplankton Community Structure in the Bering Sea," *Marine Ecology Progress Series*, v. 352 (2007), p. 14.

[22] T. F. Cooper et al., "Declining Coral Calcification in Massive Porites in Two Nearshore Regions of the Northern Great Barrier Reef," *Global Change Biology*, v. 14 (2008): 529-538.

[23] Bruce Mapstone, "Acid Oceans in the Spotlight," *Antarctic Climate and Ecosystem News*, edition 4 (August 2008): 1.

[24] M. J. Atkinson, B. Carlson, and G. L. Crow, "Coral growth in high nutrient, low-pH seawater: a case study of corals cultured at the Waikiki Aquarium, Honolulu, Hawaii," *Coral Reefs*, v. 14, no. 4 (1995): 215-223.

[25] The Royal Society, *Ocean Acidification due to Increasing Atmospheric Carbon Dioxide*, Policy Document 12/05 (June 2005), 60 p.

[26] Testimony by Dr. Richard A. Feely, NOAA Office of Oceanic and Atmospheric Research, before the House Committee on Science and Technology, Subcommittee on Energy and Environment, June 5, 2008.

[27] Alaska Fisheries Science Center, *Forecast Fish, Shellfish, and Coral Population Responses to Ocean Acidification in the North Pacific Ocean and Bering Sea*, National Marine Fisheries Service (Juneau, AK: August 2008), AFSC Processed Report 2008-7, 35 p.

[28] Personal communication from Jonathan Kelsey, Congressional Affairs Specialist, Office of Legislative Affairs, NOAA, (202) 482-0809, Dec. 11, 2008.

[29] See http://www.noaanews.noaa.gov/stories2008/20081020_oceanacid.html.

[30] This measure requires the Secretary of Commerce to request that the National Research Council study acidification of the oceans and how this process affects the United States.

[31] See http://oco.jpl.nasa.gov/, http://modis.gsfc.nasa.gov/about/, http://oceancolor.gsfc.nasa.gov/SeaWiFS/, and http://so-gasex.org/media

[32] See CRS Current Legislative Issue "Climate Change", available at http://apps.crs.gov/cli/cli.aspx? PRDS_CLI_ITEM_ID=2645&from=3&fromId=2522.

[33] For more details on ocean acidification legislation in the 110th Congress, see the section "Climate Change" in CRS Report RL33813, *Fishery, Aquaculture, and Marine Mammal Legislation in the 110th Congress*, by Eugene H. Buck.

In: Poisoning and Acidification of the Earth's ... ISBN: 978-1-60741-560-2
Editors: Geoffrey E. Mason

Chapter 4

OCEAN DUMPING ACT: A SUMMARY OF THE LAW

Claudia Copeland
Resources and Environmental Policy
Resources, Science, and Industry Division

SUMMARY

The Marine Protection, Research, and Sanctuaries Act has two basic aims: to regulate intentional ocean disposal of materials, and to authorize related research. Permit and enforcement provisions of the law are often referred to as the Ocean Dumping Act. The basic provisions of the act have remained virtually unchanged since 1972, when it was enacted to establish a comprehensive waste management system to regulate disposal or dumping of all materials into marine waters that are within U.S. jurisdiction, although a number of new authorities have been added. This report presents a summary of the law.

INTRODUCTION

The Marine Protection, Research, and Sanctuaries Act of 1972 (MPRSA, P.L. 92-532) has two basic aims: to regulate intentional ocean disposal of materials, and to authorize related research. Title I of the act, which is often referred to as the Ocean Dumping Act, contains permit and enforcement provisions for ocean dumping. Research provisions are contained in Title II; Title IV authorizes a regional marine research program; and Title V addresses coastal water quality monitoring. The third title of the MPRSA, which authorizes the establishment of marine sanctuaries, is not addressed here.

This report presents a summary of the law, describing the essence of the statute. It is an excerpt from a larger document, CRS Report RL30798, *Environmental Protection Laws: Summaries of Major Statutes Administered by the Environmental Protection Agency (EPA)*. Many details and secondary provisions are omitted here, and even some major components are only briefly mentioned. Further, this report describes the statute without discussing its implementation. **Table 1** shows the original enactment and subsequent amendments. **Table 2**, at the end of this report, cites the major U.S. Code sections of the codified statute.

Table 1. Ocean Dumping Act and Amendments (codified as 33 U.S.C. 1401-1445, 16 U.S.C. 1431-1447f, 33 U.S.C. 2801-2805)

Year	Act	Public Law Number
1972	Marine Protection, Research, and Sanctuaries Act	P.L. 92-532
1974	London Dumping Convention Implementation	P.L. 93-254
1977	Authorization of appropriations	P.L. 95-153
1980	Authorization of appropriations	P.L. 96-381
1980	Authorization of appropriations	P.L. 96-572
1982	Surface Transportation Assistance Act	P.L. 97-424
1986	Budget Reconciliation	P.L. 99-272, §§6061-6065
1986	Water Resources Development Act	P.L. 99-662, §§211, 728, 1172
1987	Water Quality Act of 1987	P.L. 100-4, §508
1988	Ocean dumping research amendments	P.L. 100-627, title I
1988	Ocean Dumping Ban Act	P.L. 100-688, title I
1988	U.S. Public Vessel Medical Waste Anti-Dumping Act of 1988	P.L. 100-688, title III
1990	Regional marine research centers	P.L. 101-593, title III
1992	National Coastal Monitoring Act	P.L. 102-567, title V
1992	Water Resources Development Act	P.L. 102-580, §§504-510

BACKGROUND

The nature of marine pollution requires that it be regulated internationally, since once a pollutant enters marine waters, it knows no boundary. Thus, a series of regional treaties and conventions pertaining to local marine pollution problems and more comprehensive international conventions providing uniform standards to control worldwide marine pollution has evolved over the last 35 years.

At the same time that key international protocols were being adopted and ratified by large number of countries worldwide (in the early 1970s), the United States enacted the MPRSA to regulate disposal of wastes in marine waters that are within U.S. jurisdiction. It utilizes a comprehensive and uniform waste management system to regulate disposal or dumping of all materials into ocean waters. Prior to 1972, U.S. marine waters had been used extensively as a convenient alternative to land-based sites for the disposal of various wastes such as sewage sludge, industrial wastes, and pipeline discharges and runoff.

The basic provisions of the act have remained virtually unchanged since 1972, but many new authorities have been added. These newer parts include (1) research responsibilities for the Environmental Protection Agency (EPA); (2) specific direction that EPA phase out the disposal of "harmful" sewage sludges and industrial wastes; (3) a ban on the ocean disposal of sewage sludge and industrial wastes by December 31, 1991; (4) inclusion of Long Island Sound within the purview of the act; and (5) inclusion of medical waste provisions. Authorizations for appropriations to support provisions of the law expired at the end of FY1997 (September 30, 1997). Authorities did not lapse, however, and Congress has continued to appropriate funds to carry out the act.

Four federal agencies have responsibilities under the Ocean Dumping Act: EPA, the U.S. Army Corps of Engineers, the National Oceanic and Atmospheric Administration (NOAA), and the Coast Guard. EPA has primary authority for regulating ocean disposal of all substances except dredged spoils, which are under the authority of the Corps of Engineers. NOAA is responsible for long-range research on the effects of human-induced changes to the marine environment, while EPA is authorized to carry out research and demonstration activities related to phasing out sewage sludge and industrial waste dumping. The Coast Guard is charged with maintaining surveillance of ocean dumping.

REGULATING OCEAN DUMPING

Title I of the MPRSA prohibits all ocean dumping, except that allowed by permits, in any ocean waters under U.S. jurisdiction, by any U.S. vessel, or by any vessel sailing from a U.S. port. Certain materials, such as high-level radioactive waste, chemical and biological warfare agents, medical waste, sewage sludge, and industrial waste, may not be dumped in the ocean. Permits for dumping of other materials, except dredge spoils, can be issued by the EPA after notice and opportunity for public hearings where the Administrator determines that such dumping will not unreasonably degrade or endanger human health, welfare, the marine environment, ecological systems, or economic potentialities. The law regulates ocean dumping within the area extending 12 nautical miles seaward from the U.S. baseline and regulates transport of material by U.S.-flagged vessels for dumping into ocean waters. EPA designates sites for ocean dumping and specifies in each permit where the material is to be disposed. EPA prepares an annual report of ocean dumping permits for material other than dredged material.[1]

In 1977, Congress amended the act to require that dumping of municipal sewage sludge or industrial wastes that unreasonably degrade the environment cease by December 1981. (However, that deadline was not achieved, and amendments passed in 1988 extended the deadline to December 1991.) In 1986, Congress directed that ocean disposal of all wastes cease at the traditional 12-mile site off the New York/New Jersey coast (that is, it barred issuance of permits at the 12-mile site) and directed that disposal be moved to a new site 106 miles offshore. In 1988, Congress enacted several laws amending the Ocean Dumping Act, with particular emphasis on phasing out sewage sludge and industrial waste disposal in the ocean, which continued despite earlier legislative efforts.

In 1992, Congress amended the act to permit states to adopt ocean dumping standards more stringent than federal standards and to require that permits conform with long-term management plans for designated dumpsites, to ensure that permitted activities are consistent with expected uses of the site.

Virtually all ocean dumping that occurs today is dredged material, sediments removed from the bottom of waterbodies in order to maintain navigation channels and berthing areas. Other materials that are dumped include vessels, fish wastes, and human remains. The Corps of Engineers issues permits for ocean dumping of dredged material, the bulk of which results from maintenance dredging by the Corps itself or its contractors.

According to data compiled by the Corps, each year an average of 70 million cubic yards of dredged sediment material is disposed of in the ocean at designated sites.[2] Before sediments can be permitted to be dumped in the ocean, they are evaluated to ensure that the dumping will not cause significant harmful effects to human health or the marine environment. EPA is responsible for developing criteria to ensure that the ocean disposal of dredge spoils does not cause environmental harm. Permits for ocean disposal of dredged material are to be based on the same criteria utilized by EPA under other provisions of the act, and to the extent possible, EPA-recommended dumping sites are used. Where the only feasible disposition of dredged material would violate the dumping criteria, the Corps can request an EPA waiver. Amendments enacted in 1992 expanded EPA's role in permitting of dredged material by authorizing EPA to impose permit conditions or even deny a permit, if necessary to prevent environmental harm.

Permits issued under the Ocean Dumping Act specify the type of material to be disposed, the amount to be transported for dumping, the location of the dumpsite, the length of time the permit is valid, and special provisions for surveillance. The EPA Administrator can require a permit applicant to provide information necessary for the review and evaluation of the application.

ENFORCEMENT

The act authorizes EPA to assess civil penalties of not more than $50,000 for each violation of a permit or permit requirement, taking into account such factors as gravity of the violation, prior violations, and demonstrations of good faith; however, no penalty can be assessed until after notice and opportunity for a hearing. Criminal penalties (including seizure and forfeiture of vessels) for knowing violations of the act also are authorized. In addition, the act authorizes penalties for ocean dumping of medical wastes (civil penalties up to $125,000 for each violation and criminal penalties up to $250,000, five years in prison, or both). The Coast Guard is directed to conduct surveillance and other appropriate enforcement activities to prevent unlawful transportation of material for dumping, or unlawful dumping. Like many other federal environmental laws, the Ocean Dumping Act allows individuals to bring a citizen suit in U.S. district court against any person, including the United States, for violation of a permit or other prohibition, limitation, or criterion issued under Title I of the act.

In conjunction with the Ocean Dumping Act, the Clean Water Act (CWA) regulates all discharges into navigable waters including the territorial seas. Although these two laws overlap in their coverage of dumping from vessels within the territorial seas, any question of conflict is essentially moot because EPA has promulgated a uniform set of standards (40 CFR Parts 220-229). The Ocean Dumping Act preempts the CWA in coastal waters or open oceans, and the CWA controls in estuaries. States are permitted to regulate ocean dumping in waters within their jurisdiction under certain circumstances.

The Ocean Dumping Act also requires the Administrator, to the extent possible, to apply the standards and criteria binding upon the United States that are stated in the 1972 Convention on the Prevention of Marine Pollution by Dumping of Wastes and Other Matters (known as the London Dumping Convention). This convention, signed by more than 85 countries, includes annexes that prohibit the dumping of mercury, cadmium and other substances such as DDT and PCBs, solid wastes and persistent plastics, oil, high- level radioactive wastes, and chemical and biological warfare agents; and requires special permits for other heavy metals, cyanides and fluorides, and medium- and low-level radioactive wastes.

RESEARCH AND COASTAL WATER QUALITY MONITORING

Title II of the MPRSA authorizes two types of research: general research on ocean resources, under the jurisdiction of NOAA; and EPA research related to phasing out ocean disposal activities.

NOAA is directed to carry out a comprehensive, long-term research program on the effects not only of ocean dumping, but also of pollution, overfishing, and other human-induced changes on the marine ecosystem. Additionally, NOAA assesses damages from spills of petroleum and petroleum products.

EPA's research role includes "research, investigations, experiments, training, demonstrations, surveys, and studies" to minimize or end the dumping of sewage sludge and industrial wastes, along with research on alternatives to ocean disposal. Amendments in 1980 required EPA to study technological options for removing heavy metals and certain organic materials from New York City's sewage sludge.

Table 2. Major U.S. Code Sections of the Marine Protection, Research, and Sanctuaries Act (codified as 33 U.S.C. 1401-1445, 16 U.S.C. 1431-1447f, 33 U.S.C. 2801-2805)

	Section Title	Ocean Dumping Act
33 U.S.C.		
1401	Congressional findings, declaration of policy	Sec. 2
1401	Definitions	Sec. 3
Title I -	**Permit Program**	
1411	Prohibited acts	Sec. 101
1412	Environmental Protection Agency permits	Sec. 102
1413	Corps of Engineers permits	Sec. 103
1414	Permit conditions	Sec. 104
1414a	Special provisions regarding certain dumping sites	Sec. 104A
1414b	Ocean dumping of sewage sludge and industrial waste	Sec. 104B
1414c	Prohibition on disposal of sewage sludge at landfills on Staten Island	Sec. 104C
1415	Penalties	Sec. 105
1416	Relationship to other laws	Sec. 106
1417	Enforcement	Sec. 107
1418	Regulations	Sec. 108
1419	International cooperation	Sec. 109
1420	Authorization of appropriations	Sec. 111
1421	Annual report to Congress	Sec. 112
Title II -	**Research Programs**	
1441	Monitoring and research programs	Sec. 201
1442	Research on long-term effects	Sec. 202
1443	Research program - ocean dumping and other methods	Sec. 203
1444	Annual reports	Sec. 204
1445	Authorization of appropriations	Sec. 205
Title III	**Marine Sanctuaries**	
Title IV - 16 U.S.C.	**Regional Marine Research Programs**	
1447	Purposes	Sec. 401
1447a	Definitions	Sec. 402

Table 2. (Continued)

	Section Title	Ocean Dumping Act
1447b	Regional marine research boards	Sec. 403
1447c	Regional research plans	Sec. 404
1447d	Research grant program	Sec. 405
1447e	Report on research program	Sec. 406
1447f	Authorization of appropriations	Sec. 407
Title V - 33 U.S.C.	**National Coastal Monitoring System**	
2801	Purposes	Sec. 501
2802	Definitions	Sec. 502
2803	Comprehensive coastal water quality monitoring program	Sec. 503
2804	Report to Congress	Sec. 504
2805	Authorization of appropriations	Sec. 505

Note: This table shows the major code sections. For more detail and to determine when a section was added, the reader should consult the printed version of the U.S. Code.

Title IV of the MPRSA established nine regional marine research boards for the purpose of developing comprehensive marine research plans, considering water quality and ecosystem conditions and research and monitoring priorities and objectives in each region. The plans, after approval by NOAA and EPA, are to guide NOAA in awarding research grant funds under this title of the act.

Title V of the MPRSA established a national coastal water quality monitoring program. It directs EPA and NOAA jointly to implement a long-term program to collect and analyze scientific data on the environmental quality of coastal ecosystems, including ambient water quality, health and quality of living resources, sources of environmental degradation, and data on trends. Results of these activities (including intensive monitoring of key coastal waters) are intended to provide information necessary to design and implement effective programs under the Clean Water Act and Coastal Zone Management Act.

End Notes

[1] See [http://www.epa.gov/owow/oceans

[2] See [http://el.erdc.usace.army.mil/odd/default.htm].

In: Poisoning and Acidification of the Earth's... ISBN: 978-1-60741-560-2
Editors: Geoffrey E. Mason

Chapter 5

OIL SPILLS IN U.S. COASTAL WATERS: BACKGROUND, GOVERNANCE, AND ISSUES FOR CONGRESS

Jonathan L. Ramseur
Environmental Policy Resources, Science, and Industry Division

SUMMARY

During the past two decades, while U.S. oil imports and consumption have steadily risen, oil spill incidents and the volume of oil spilled have not followed a similar course. In general, the annual number and volume of oil spills have shown declines — in some cases, dramatic declines. The 1989 *Exxon Valdez* spill in Alaskan waters played a large role in stimulating actions that contributed to this trend, particularly the decrease in the annual spill volumes. The *Exxon Valdez* spill highlighted the need for stronger legislation, inflamed public sentiment, and spurred Congress to enact comprehensive oil spill legislation, resulting in the Oil Pollution Act of 1990 (P.L. 101-380). This law expanded and clarified the authority of the federal government and created new oil spill prevention and preparedness requirements. Moreover, the 1990 legislation strengthened existing liability provisions, providing a greater deterrent against spills. After 1990, spill volume from oil tankers, the vessels that carry and have spilled the most oil, decreased significantly.

Considering that U.S. oil consumption and oil imports have steadily increased, the trend of declining spill incidents and volume in past years is noteworthy. Yet, recent annual data indicate that the overall decline of annual spill events may have stopped. Although Energy Information Administration projections indicate that oil imports are expected to decline in coming years, the United States is projected to continue importing more than 50% of the oil it consumes. The threat of oil spills raises the question of whether U.S. officials have the necessary resources at hand to respond to a major spill. There is some concern that the favorable U.S. spill record has resulted in a loss of experienced personnel, capable of responding quickly and effectively to a major oil spill. Moreover, the level of funding required to respond to such a spill, particularly its aftermath, may be currently inadequate, according to U.S. Coast Guard reports.

Recent oil spills into San Francisco Bay (November 2007) and the Mississippi River (July 2008) have generated further interest and debate concerning oil spill policy issues, such as oil spill response and prevention.

No oil spill is entirely benign. Even a relatively minor spill, depending on the timing and location, can cause significant harm to individual organisms and entire populations. Marine mammals, birds, bottom-dwelling and intertidal species, and organisms in early developmental stages — eggs or larvae — are especially vulnerable to a nearby spill. However, the effects of oil spills can vary greatly. Oil spills can cause impacts over a range of time scales, from only a few days to several years, or even decades in some cases.

This report reviews the history and trends of oil spills in the United States; identifies the legal authorities governing oil spill prevention, response, and cleanup; and examines the threats of future oil spills in U.S. coastal waters.

INTRODUCTION

Oil is the dominant source of energy in the United States, supplying the nation with approximately 40% of its energy needs. Its use is widespread, providing fuel for the transportation, industrial, and residential sectors. Vast quantities of oil continuously enter the country via vessel or pipeline and are then transported to destinations throughout the nation. With such widespread use and nonstop movement, it is inevitable that some number of spills will occur. One continuing policy issue is whether the nation has the necessary resources and personnel in place to respond to a major spill.

Several major U.S. oil spills have had lasting repercussions that transcended the local environmental and economic effects. The most notable example is the 1989 *Exxon Valdez* spill, which released approximately 11 million gallons of crude oil into Prince William Sound, Alaska. The *Exxon Valdez* spill — the largest and most expensive oil spill in U.S. waters to date[1] — produced extensive consequences beyond Alaska. According to the National Academies of Science, the *Exxon Valdez* disaster caused "fundamental changes in the way the U.S. public thought about oil, the oil industry, and the transport of petroleum products by tankers ... 'big oil' was suddenly seen as a necessary evil, something to be feared and mistrusted."[2]

This report focuses on oil spills[3] in U.S. coastal waters.[4] The first section highlights background issues, including oil spill statistics and potential environmental impacts. The second section discusses the legal framework that governs oil spill prevention and response. The third section examines the threat of future oil spills in coastal waters and whether response personnel are prepared to respond to a major spill. The final section highlights recent legislative activity.

BACKGROUND

Oil Spills in U.S. Coastal Waters

While U.S. oil imports and consumption have steadily risen, oil spill incidents and volume spilled have not followed a similar course. (See **Figure 1**.) In general, oil spill events and the volume of oil released have declined over the past two decades; in some years, the declines have been dramatic. Note that this figure does not include data from 2005, described in the text box below. The decline of *spill incidents* is likely related, at least in part, to international oil pollution standards that went into effect in 1983. These new standards were implemented in the United States by the Act to Prevent Pollution from Ships.[5]

The substantial drop in the annual *spill volume* is most attributable to the decline in volume spilled by oil tankers and barges — the vessels that transport oil and have historically spilled the most oil. As shown in **Figure 2,** the volume of oil spilled from vessels in U.S. waters in the 1990s differed dramatically from the volume spilled in the 1 980s. The *Exxon Valdez* spill of 1989 and the resulting Oil Pollution Act of 1990 (OPA) played key roles in the subsequent spill volume reduction. The 1990 Act (discussed below) made comprehensive changes to U.S. oil pollution law by expanding federal response authority and increasing spill liability. The high costs associated with the *Exxon Valdez* spill,[6] and the threat of broad liability imposed by OPA (in some scenarios, unlimited liability), have likely been the central drivers for the spill volume decline seen in the 1 990s.

SPILLS ASSOCIATED WITH HURRICANES KATRINA AND RITA (2005)

As of the date of this report, the Coast Guard database did not include 2005 spill data. However, there was a substantial increase in spill volume in 2005, due to spills associated with Hurricanes Katrina and Rita. The Minerals Management Service (MMS) summary of hurricane incidents includes the following:

- More than 8 million gallons of oil from aboveground oil storage facilities; however, 50% was reported recovered, and it is unknown has much of the remaining volume impacted coastal waters;
- Over 600,000 gallons (including an estimated 84,000 gallons from one platform incident) were spilled from federal offshore oil platforms and associated pipelines;
- Approximately 3.3 million gallons were spilled from a tank barge, when it struck a submerged oil platform that had been damaged during the storms.

MMS, *Petroleum Spills of One Barrel and Greater from Federal Outer Continental Shelf Facilities Resulting from Damages Caused by 2005 Hurricanes Katrina and Rita Including Post-Hurricane Seepage through December 2007* (revised June 23, 2008), at [http://www.mms.gov/incidents/].

In addition to international and federal governance, 28 states had oil spill liability laws, 19 of which imposed unlimited liability, before the *Exxon Valdez* spill occurred in 1989.[7] After the 1989 spill, some states enacted additional legislation,[8] which may have contributed to the declines.

The **Appendix** to this report contains additional information, including a further breakdown of oil inputs in coastal waters by source category. The

Appendix also provides oil spill data and analysis specific to onshore facilities and pipelines, as well as offshore oil extraction operations.

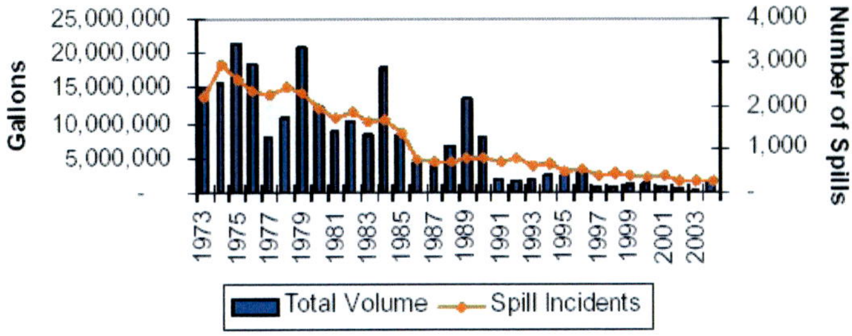

Source: Prepared by CRS with data from the United States Coast Guard (USCG) Oil Spill Compendium, available at [http://www.uscg.mil/hq/g-cp/comrel/factfile/index.htm].

Figure 1. Volume and Number of Oil Spills for Incidents Above 100 Gallons in U.S. Coastal Waters, 1973-2004

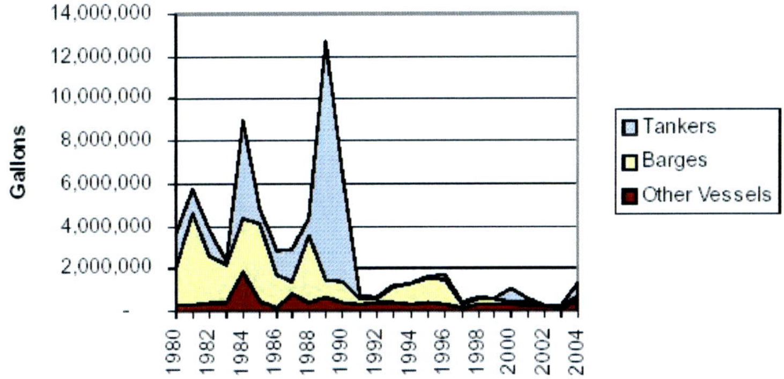

Source: Prepared by CRS with data from the USCG Oil Spill Compendium.
Figure 2. Volume of Oil Spilled from Vessels into U.S. Coastal Waters, 1980-2004

Impacts of Oil Spills in Aquatic Environments

No oil spill is entirely benign. Depending on timing and location, even a relatively minor spill can cause significant harm to individual organisms and entire populations.[9] Oil spills can cause impacts over a range of time scales, from days to years, or even decades for certain spills. Impacts are typically divided into acute (short-term) and chronic (long-term) effects. Both types are part of a complicated and often controversial equation that is addressed after an oil spill: ecosystem recovery.

Acute Impacts

Depending on the toxicity and concentration of the spill, acute exposure to oil spills can kill various organisms and cause the following debilitating (but not necessarily lethal) effects:[10]

- reduced reproduction,
- altered development,
- impaired feeding mechanisms, and
- decreased defense from disease.

Birds, marine mammals, bottom-dwelling and intertidal species, and organisms in their developmental stages — e.g., fish eggs and larvae — are particularly vulnerable to oil spills.[11]

In addition to the impacts to individual organisms, oil spills can lead to a disruption of the structure and function of the ecosystem. Certain habitats — such as coral reefs, mangrove swamps, and salt marshes — are especially vulnerable, because the physical structure of the habitats depends upon living organisms.

These potential acute effects to individual organisms and marine ecosystems have been "unambiguously established" by laboratory studies and well-studied spills, such as the *Exxon Valdez*.[12]

Chronic Impacts

Long-term, chronic exposure typically occurs from continuous oil releases — leaking pipelines, offshore production discharges, and non- point sources (e.g., urban runoff). Although spills are normally associated with acute impacts, some oil spills have also demonstrated chronic exposure and effects.[13] There is increasing evidence that chronic, low-level exposures to oil contaminants can significantly affect the survival and reproductive success of marine birds and mammals.[14] However, because of the complexity of factors, including a longer time period and presence of other pollutants, determining the precise effects on species and ecosystems due to chronic oil exposure in a particular locale is difficult for scientists. As a result, studies involving chronic effects are often met with debate and some controversy.

Ecosystem Recovery

Interested parties may have differing opinions at to what constitutes ecosystem recovery. At one end of the spectrum, local groups may demand that an ecosystem be returned to pre-spill conditions. NOAA regulations (15 CFR Section 990.30) state that recovery "means the return of injured natural resources and services to baseline" — in other words, a return to conditions as they would have been had the spill not occurred. Baseline conditions may not equate with pre-spill conditions. Multiple variables affect local species and ecosystem services. For example, one species at a spill site could have been on the decline at the time of an incident, because of changing water temperatures, for example. These types of trends are considered when trustees evaluate restoration efforts. Restoration leaves room for site-specific interpretation, which, in the case of the *Exxon Valdez* spill and cleanup, continues to generate considerable argument.

Economic Costs of Oil Spills

The economic costs that can result from an oil spill can be broken into three categories: cleanup expenses, natural resource damages, and the various economic losses incurred by the affected community or individuals.

Cleanup Costs

The cleanup costs of an oil spill can vary greatly and are influenced by a mix of factors: location characteristics, oil type, and oil volume. Location is generally considered the most important factor because it involves multiple variables. Areas with less water movement, such as marshlands, will generally cost more to clean up than open water. Tourist destinations or sensitive habitats, such as coral reefs, will likely require more stringent cleanup standards, thus increasing the costs. The political and social culture at the spill site plays a part as well. A spill in a high-profile area — e.g., the November 7, 2007 spill (53,000 gallons) from a container ship into the San Francisco Bay — will likely receive special attention. Major oil spills, especially ones that affect shoreline ecosystems, are often met with extensive media coverage, placing pressure on parties to take action. Coupled with this pressure, authorities (federal or state) at these locations may require extensive oil spill response requirements, which can influence cleanup cost. The United States likely meets this description, because its average cleanup cost (per barrel of oil spilled) is considerably higher than in other parts of the world.[15]

The more persistent and viscous oil types, such as heavy crude oil and intermediates known as bunker fuels, are more expensive to clean up. Gasoline and other lighter refined products may require only minimal cleanup action. These materials will evaporate or disperse relatively quickly, leaving only a small volume of petroleum product available for recovery.

Compared with other factors, spill volume is less important. A major spill away from shore will likely cost considerably less than a minor spill in a sensitive location. Certainly, the amount of oil spilled affects cleanup costs, because, all things being equal, a larger spill will require a larger and more expensive cleanup effort. However, the relationship between cleanup costs and spill volume is not linear. Cleaning up a smaller spill is likely to cost more than a larger spill on a per-gallon basis.[16]

Natural Resources Damages

This category of costs relates to the environmental impacts caused by an oil spill. Pursuant to OPA, the party responsible for an oil spill is liable for any loss of natural resources (e.g., fish, animals, plants, and their habitats) and the services provided by the resource (e.g., drinking water, recreation).

When a spill occurs, natural resource trustees conduct a natural resource damage assessment to determine the extent of the harm. Trustees may include representatives from tribal governments as well as officials from state agencies (designated by the relevant Governor) and federal agencies (designated by the President), such as NOAA.[17]

The Oil Pollution Act (OPA) of 1990 states that the measure of natural resource damages includes

- the cost of restoring, rehabilitating, replacing, or acquiring the equivalent of, the damaged natural resources;
- the diminution in value of those natural resources pending restoration; and
- the reasonable cost of assessing those damages.[18]

Pursuant to OPA, NOAA developed regulations pertaining to natural resource damage assessments in 1996.[19] Natural resource damages may include both losses of direct use and passive uses. Direct use value may derive from recreational (e.g., boating), commercial (e.g., fishing), or cultural or historical uses of the resource. In contrast, a passive-use value may derive from preserving the resource for its own sake or for enjoyment by future generations.[20]

The damages are compensatory, not punitive. Collected damages cannot be placed into the general treasury revenues of the federal or state government, but must be used to restore or replace lost resources.[21] Indeed, NOAA's regulations focus on the costs of primary restoration — returning the resource to its baseline condition — and compensatory restoration — addressing interim losses of resources and their services.[22]

Other Economic Costs

Oil spills can generate costs other than response expenses or damages to natural resources. An oil spill can disrupt business activity near the spill, particularly businesses that count on the reputation of the local environment. For example, the local tourist industry may be affected. In some cases, a well-publicized oil spill can weaken the tourist industry near the spill site, regardless of the actual threat to human health created by the spill.

Local infrastructure and services can be disrupted by an oil spill. Port and harbor operations may be interrupted, altering the flow of trade goods. Power plants that use cooling water systems may need to temporarily cease operations. For example, the Salem Nuclear Plant — the second largest nuclear plant in the United States — was forced to halt activity due to a substantial oil spill (more than 250,000 gallons) in the Delaware River in November 2004.

Unlike natural resource damage claims, which are brought by the appropriate natural resource trustees, the costs described in this section would be submitted as claims by the third-parties suffering the specific loss.

OIL SPILL GOVERNANCE

When the *Exxon Valdez* ran aground in March 1989, there were multiple federal statutes, state statutes, and international conventions that dealt with oil discharges. The governing framework for oil spills in the United States remains a combination of federal, state, and international authorities. Within this framework, several federal agencies have the authority to implement oil spill regulations. The framework and primary federal funding process (the Oil Spill Liability Trust Fund) used to respond to oil spills are described below.

Federal Authorities: Before and After the *Exxon Valdez* Spill

The following list highlights the primary federal authorities that were in effect when the *Exxon Valdez* spill occurred in 1989:

- Clean Water Act (1972):[23] The Clean Water Act (CWA) represented the broadest authority for addressing oil spills at the time of the *Exxon Valdez* spill. Section 311 of the CWA established requirements for oil spill reporting, response, and liability. The act also created a fund (311 Fund), maintained by federal appropriations, that could be used for cleanup and natural resource restoration.
- Deepwater Port Act (1 974):[24] This statute addressed oil spills and liability issues at deepwater oil ports. The act also set up the Deepwater Port Fund to provide for prompt cleanup and compensate damages above liability limits. The fund was financed by a per- gallon tax on oil transferred at a deepwater port.
- Trans-Alaska Pipeline Authorization Act (1973):[25] This act covered oil spills and liability relating to the Trans-Alaska Pipeline System (TAPS). Although the pipeline is constructed over land, spills from it could reach coastal waters via inland rivers. The act created a trust fund, financed through a lessee fee, that could be used to respond to spills and damages from the pipeline.
- Outer Continental Shelf Lands Act Amendments (1978):[26] This act established an oil spill liability structure and rules for oil extraction facilities in federal offshore waters. With this legislation, Congress created the Offshore Pollution Fund, financed by a per-gallon fee on produced oil, that could be used for oil spill cleanup and damages.
- National Oil and Hazardous Substances Pollution Contingency Plan (NCP): The NCP was first established in 1968, after U.S. policymakers observed the response to a 37-million-gallon oil tanker spill (*Torrey Canyon*) off the coast of England.[27] The NCP contains the federal government's procedures for responding to oil spills and hazardous substance releases.[28] Subsequent laws have amended the NCP, including the CWA in 1972 and the Comprehensive Environmental Response, Compensation, and Liability Act (CERCLA or Superfund) in 1980.

After the *Exxon Valdez* spill, many observers[29] described the above legal collection as an ineffective patchwork. Arguably, each law had perceived shortcomings (discussed below in the context of post-*Exxon Valdez* legislation), and none provided comprehensive oil spill coverage.

For more than 15 years prior to the *Exxon Valdez* incident, Congress made attempts to enact a unified oil pollution law. Several contentious issues produced deadlocks, hindering the passage of legislation. One of the central points of debate, state preemption, dealt with whether a federal oil spill law should limit a state's ability to impose stricter requirements, particularly unlimited liability. Other liability questions also generated debate. For example, if an oil spill occurred, should the owner of the cargo (i.e., oil) be held liable, as was the ship owner/operator? Another point of contention was whether oil-carrying vessels should be required to have double hulls. Although proponents argued that a second hull would help prevent oil spills, the shipping industry raised concern that implementing such a mandate would disrupt oil transportation and potentially affect the national economy. A final issue involved the interaction between domestic legislation (federal and state) and international measures. Some were concerned that if the United States became a party to certain international agreements under consideration in the 1980s,[30] the international standards would preempt federal and state laws, especially those establishing liability limits. Proponents argued that these concerns were overstated, and stressed that joining the international agreements was especially important for the United States because of the international nature of oil transportation and associated pollution.

Following the 1989 *Exxon Valdez* spill, Members faced great pressure to overcome the disputes discussed above.[31] The spill highlighted the inadequacies of the existing coverage and generated public outrage. The end result was the Oil Pollution Act of 1990 (OPA)[32] — the first comprehensive law to specifically address oil pollution to waterways and coastlines of the United States.

Oil Pollution Act of 1990

With the enactment of OPA on August 18, 1990, Congress consolidated the existing federal oil spill laws under one program. The 1990 law expanded the existing liability provisions within the CWA and created new free-standing requirements regarding oil spill prevention and response. Key OPA provisions are discussed below.

Spill Response Authority

When responding to a spill, many considered the lines of responsibility under the pre-OPA regime to be unclear,[33] with too much reliance on spillers to perform proper cleanup.[34] OPA strengthened and clarified the federal government's role in oil spill response and cleanup. OPA Section 4201 amended Section 311(c) of the CWA to provide the President (delegated to the USCG or EPA) with three options: perform cleanup immediately ("federalize" the spill), monitor the response efforts of the spiller, or direct the spiller's cleanup activities. The revised response authorities addressed concerns "that precious time would be lost while waiting for the spiller to marshall its cleanup forces."[35]

The federal government determines the level of cleanup required. Although the federal government must consult with designated trustees of natural resources and the governor of the state affected by the spill, the decision that cleanup is completed and can be ended rests with the federal government. States may require further work, but without the support of federal funding.[36]

National Contingency Plan

OPA expanded the role and breadth of the NCP. The 1990 law established a multi-layered planning and response system to improve preparedness and response to spills in marine environments.[37] Among other things, the act also required the President to establish procedures and standards (as part of the NCP) for responding to worst-case oil spill scenarios.[38]

Tank Vessel and Facility Response Plans

As a component of the enhanced NCP, OPA amended the CWA to require that U.S. tank vessels, offshore facilities, and certain onshore facilities[39] prepare and submit oil spill response plans to the relevant federal agency. In general, vessels and facilities are prohibited from handling, storing, or transporting oil if they do not have a plan approved by (or submitted to) the appropriate agency[40] (**Table 1**).

The plans should, among other things, identify how the owner or operator of a vessel or facility would respond to a worst-case scenario spill. Congress did not intend for every vessel to have onboard all the personnel and equipment needed to respond to a worst-case spill, but vessels must have a plan and procedures to call upon — typically through a contractual relationship — the necessary equipment and personnel for responding to a worst-case spill.[41]

In 2004, Congress enacted an amendment requiring non-tank vessels (i.e., ships carrying oil for their own fuel use) over 400 gross tons to prepare and submit a vessel response plan.[42] Congress reasoned that many non-tank vessels have as much oil onboard as small tank vessels, thus presenting a comparable risk from an oil spill. Moreover, the international standards for oil spill prevention[43] apply to tanker and non-tanker vessels alike. Thus, the 2004 amendment brought the U.S. law more in line with international provisions.

Double-Hull Design for Vessels

The issue of double hulls received considerable debate for many years prior to OPA, and it was one of the stumbling blocks for unified oil spill legislation. Proponents maintained that double-hull construction provides extra protection if a vessel becomes damaged.[44] However, opponents argued that a double-hulled vessel might cause stability problems if an accident occurred, thus negating the benefits.[45] Stakeholders also highlighted the impacts that a double-hull requirement would entail for the shipping industry (e.g., cost and time of retrofitting, ship availability).[46] The OPA requirements for double hulls reflected some of these concerns.

The act required new vessels carrying oil and operating in U.S. waters to have double hulls.[47] However, OPA provided certain exceptions, depending on the size of the vessel (e.g., less than 5,000 gross tons)[48] and its particular use (e.g., lightering).[49] For older vessels, OPA established a staggered retrofitting schedule, based on vessel age and size. Many of the age-based deadlines have already passed. By 2015 at the latest, the law requires that all oil-carrying vessels operating in U.S. waters have double hulls.

Liability Issues

OPA unified the liability provisions of existing oil spill law, creating a freestanding liability regime. Section 1002 states that responsible parties are liable for any discharge of oil (or threat of discharge) from a vessel or facility[50] to navigable waters, adjoining shorelines, or the exclusive economic zone[51] of the United States (i.e., 200 miles beyond the shore).

Regarding the existing oil spill law prior to OPA, Congress recognized that "there is no comprehensive legislation in place that promptly and adequately compensates those who suffer other types of economic loss as a result of an oil pollution incident."[52] OPA broadened the scope of damages (i.e., costs) for which an oil spiller would be liable. Under OPA, a responsible party is liable for all cleanup costs incurred, not only by a government entity, but also by a private party.[53] In addition to cleanup costs, OPA significantly increased the range of liable damages to include the following:

- injury to natural resources,
- loss of personal property (and resultant economic losses),
- loss of subsistence use of natural resources,
- lost revenues resulting from destruction of property or natural resource injury,
- lost profits resulting from property loss or natural resource injury, and
- costs of providing extra public services during or after spill response.[54]

OPA provided several defenses from liability: act of God, act of war, and act or omission of a third party. Although these defenses are more narrow than those for oil spills under the pre-OPA framework (primarily the CWA), they are similar to those of the Superfund statute,[55] established in 1980 for releases of hazardous substances.

Except for certain behavior, including acts of gross negligence or willful misconduct,[56] OPA set liability limits (or caps) for cleanup costs and other damages. Liability limits for vessels are based on vessel carrying capacity, generally $1,200 per gross ton. As an example, the liability limit for the 2004 *Athos* tanker spill in Delaware River was approximately $45 million.[57] Offshore facility liability is capped at $75 million; onshore facilities and deepwater port liability is limited to $350 million. These limits are much higher than under the pre-OPA liability structure.

OPA required the President to issue regulations to adjust the liability limits at least every three years to take into account changes in the consumer price index (CPI). Despite this requirement, adjustments to liability limits were not made until July 2006. The Coast Guard and Maritime Transportation Act of 2006 (P.L. 109- 241) amended OPA, increasing limits to $1,900/gross ton for double-hulled vessels and $3,000/gross ton for single-hulled vessels.

The Oil Spill Liability Trust Fund

Prior to OPA, federal funding for oil spill response was generally considered inadequate,[58] and damage recovery was difficult for private parties.[59] To help address these issues, Congress established the Oil Spill Liability Trust Fund (OSLTF).

Pursuant to Executive Order (EO) 12777, the USCG created the National Pollution Funds Center (NPFC) to manage the trust fund in 1991. The fund may be used for several purposes:

- prompt payment of costs for responding to and removing oil spills;
- payment of the costs incurred by the federal and state trustees of natural resources for assessing the injuries to natural resources caused by an oil spill, and developing and implementing the plans to restore or replace the injured natural resources;
- payment of parties' claims for uncompensated removal costs, and for uncompensated damages (e.g., financial losses of fishermen, hotels, and beachfront businesses);
- payment for the net loss of government revenue, and for increased public services by a state or its political subdivisions; and
- payment of federal administrative and operational costs, including research and development, and $25 million per year for the Coast Guard's operating expenses.

Although Congress created the OSLTF in 1986,[60] Congress did not authorize its use or provide its funding until after the *Exxon Valdez* incident. In 1990, OPA provided the statutory authorization necessary to put the fund in motion. Through OPA, Congress transferred other federal liability funds[61] into the OSLTF. In complementary legislation, Congress imposed a 5-cent-per-barrel tax on the oil industry to support the fund.[62] Collection of this fee ceased on December 31, 1994, due to a sunset provision in the law. However, in April 2006, the tax resumed as required by the Energy Policy Act of 2005 (P.L. 109-58). The level of funding in the trust fund is discussed below.

Financial Responsibility

To preserve the trust fund and ensure that responsible parties can be held accountable for oil spill cleanup and damages, OPA requires that vessels maintain evidence of financial responsibility (e.g., insurance). The Coast Guard's National Pollution Funds Center carries out this mandate by issuing Certificates of Financial Responsibility (COFRs) to shipping vessel owners when owners demonstrate the ability to pay for oil spill cleanup and damages.

In general, vessels over 300 gross tons are required to have a valid COFR to operate in U.S. waters.

Although the liability limits were increased with the Coast Guard and Maritime Transportation Act of 2006 (P.L. 109-241), the Coast Guard regulations (33 CFR Part 138) addressing COFRs have not been amended to reflect the new limits. In a *Federal Register* notice (Vol. 71, August 18, 2006) the Coast Guard stated that "the levels of financial responsibility enforceable by the Coast Guard are the total applicable amounts currently found at 33 CFR 138.80(f)." The same *Federal Register* notice indicates that the Coast Guard will initiate a rulemaking to amend the levels of financial responsibility.

Other Federal Laws

Although OPA is the primary domestic legislation for oil spills, other federal laws contain provisions that relate to oil spills. Many of these provisions were in place before OPA. The following list is not all-inclusive, but it highlights the main requirements authorized by laws other than OPA.

Clean Water Act

Section 311(j)(1) of the 1972 CWA called for regulations to prevent the discharge of oil from vessels, onshore facilities, and offshore facilities. Pursuant to this statutory requirement,[63] the EPA crafted regulations[64] for spill prevention control and countermeasure (SPCC) plans in 1973, some of which are scheduled to go into effect July 1, 2009.[65] SPCC plans address the "procedures, methods, and equipment and other requirements for equipment to prevent discharges."[66] The EPA's SPCC plans apply only to non-transportation, onshore facilities that exceed a certain oil storage capacity and that, in the event of a spill, can be reasonably expected, because of their location, to produce an oil discharge that would reach navigable waters or adjoining shorelines of the United States.[67] Unlike other oil spill preparedness provisions, SPCC plans focus more on prevention than on response activities, requiring, for example, secondary containment (e.g., dikes, berms) for oil-storage equipment.

Outer Continental Shelf Lands Act

The primary federal law governing oil development and operations in federal waters is the Outer Continental Shelf Lands Act (OCSLA) of 1953 and its subsequent amendments. The OCSLA provided the foundation for regulations (30 CFR Part 250) that are implemented by the Minerals

Management Service (MMS). Sections of the MMS regulations address oil spill prevention and response issues by requiring that various equipment and procedures be in place at offshore facilities. For more information, see CRS Report RL3 3404, *Offshore Oil and Gas Development: Legal Framework*, by Adam Vann.

Pipeline Legislation

The U.S. pipeline network is extensive. Recent estimates indicate there are more than 33,000 miles of pipelines just in the Gulf of Mexico.[68] Moreover, U.S. inland pipelines are concentrated in coastal areas, particularly in the Gulf states, and these pipelines may have an impact on coastal waters if spills reach waterways that empty into coastal waters.

Several laws govern oil pipelines. The Hazardous Liquid Pipeline Act of 1979 (P.L. 96-129) granted authority to the Department of Transportation (DOT) to regulate various issues regarding oil spills from pipelines. On December 29, 2006, the President signed the Pipeline Safety Improvement Act of 2006 (P.L. 109-468) to improve pipeline safety and security practices, and to reauthorize the federal Office of Pipeline Safety. The Office of Pipeline Safety (OPS), which is part of the DOT, implements provisions concerning pipeline design, construction, operation and maintenance, and spill response planning. For further information on pipeline legislation, see CRS Report RL33347, *Pipeline Safety and Security: Federal Programs*, by Paul W. Parfomak.

Vessel Legislation

Several federal laws directly or indirectly deal with oil pollution from vessels.[69] Laws concerning navigation reduce the possibilities of vessel collision or hull breach by objects in the waterways.[70] Other laws call for particular vessel design standards. For example, the Ports and Waterways Safety Act of 1 972,[71] amended by the Port and Tanker Safety Act of 1978,[72] called for specific construction and equipment design requirements for oil tankers. (As noted, OPA subsequently amended this statute in 1990 by establishing a phased-in schedule for double-hulled tankers.) Congress enacted the 1970s legislation to coincide with international initiatives. In fact, many of the federal laws concerning vessel standards and pollution control procedures were written to implement international conventions. These laws are discussed in the next section.

International Conventions

The relationship between international and domestic law can be complex. In general, international conventions (i.e., treaties), when signed by the United States and (if necessary) ratified by the Senate, are on equal footing with federal law. Parties to such conventions must often implement domestic legislation to carry out the provisions outlined in the convention. Several of the federal laws governing oil spills were fashioned in this manner.[73]

International conventions have played an important role in developing consistent standards for oil-carrying vessels from different nations. A primary player in this regard is the International Maritime Organization (IMO), a body of the United Nations, which sets international maritime vessel safety and marine pollution standards. The Coast Guard represents the United States at IMO meetings.

Multiple international conventions concern vessels and their impact on the marine environment. The two conventions described below are the most relevant to oil pollution in coastal waters.

MARPOL 73/78

The IMO implements the 1973 International Convention for the Prevention of Pollution from Ships, as modified by the Protocol of 1978 (MARPOL 73/78).[74] Vessels whose nations are signatories to MARPOL are subject to its requirements, regardless of where they sail, and member nations are responsible for the vessels registered under their flag.

MARPOL 73/78 includes six annexes, each covering a different pollution type. Annex I (Prevention of Pollution by Oil) entered into force in 1983[75] and established requirements for controlling oil discharges to sea. Annex I requires vessels to have equipment that minimizes oil discharge, such as oil-water separators, and shipboard oil pollution emergency plans (SOPEPs). Although the SOPEP applicability is similar to that of the vessel response plan (VRP) required by OPA,[76] the purpose of the SOPEP is somewhat different. A SOPEP is intended to provide guidance to the vessel's officers regarding proper onboard emergency procedures when an oil spill occurs,[77] whereas the VRP is more focused on responding to the spill itself.

The United States implements Annex I through the act to Prevent Pollution from Ships (APPS).[78] APPS applies to all U.S.-flagged ships, irrespective of location, and to all foreign-flagged vessels in U.S. waters or at ports under U.S. jurisdiction. The USCG issues and enforces regulations necessary to carry out the APPS provisions. The USCG inspection program is a key component of its oil spill prevention effort.

Intervention Convention

The 1967 *Torrey Canyon* spill off the coast of Great Britain was one of the first major spills to receive worldwide attention.[79] The incident raised many questions regarding oil spill response, particularly when dealing with vessels from other nations. For example, the incident prompted debate over actions allowable if a nation's waters and environment are threatened by a spill from another nation's vessel. The 1969 International Convention Relating to Intervention on the High Seas in Cases of Oil Pollution Casualties (the Intervention Convention) sought to address these issues.

To implement this convention in the United States, Congress passed the Intervention on the High Seas Act of 1974.[80] Under this act, if the USCG determines there to be a "grave and imminent danger to the coastline or related interests of the United States from pollution or threat of pollution of the sea by convention oil [i.e., as defined in the convention]," the USCG can take action to "prevent, mitigate, or eliminate that danger."

Federal Agencies Responsibilities

Unlike most countries, the United States shares jurisdiction over its territorial seas (0-12 nautical miles from shore) with its coastal states. The 1953 Submerged Lands Act (SLA) gave coastal states jurisdiction over the submerged lands, waters, and natural resources (e.g., oil deposits) located, in most cases, within three nautical miles off the coastline.[81] The waters, seabed, and natural resources beyond the states' waters are exclusively federal, and extend to the edge of the exclusive economic zone (200 nautical miles from shore). However, the federal government maintains the authority to regulate commerce, navigation, national defense, power production, and international affairs within state waters.

The oil spill legal framework involves implementation by multiple federal agencies. Agency responsibilities can be divided into two categories: (1) oil spill response and cleanup and (2) oil spill prevention/preparedness.

Response

Oil spill response authority is determined by the location of the spill: the USCG has response authority in coastal waters, and the EPA covers inland oil spills.[82]

As the primary response authority in coastal waters, the USCG has the ultimate authority to ensure that an oil spill is effectively removed and actions are taken to prevent further discharge from the source. During response operations, the USCG coordinates the efforts of federal, state, and private parties.

Coast Guard response efforts are supported by NOAA's Office of Response and Restoration. NOAA provides scientific analysis and consultation during oil spill response activities.[83] Assistance can include oil spill tracking, cleanup alternatives, and knowledge of at-risk natural resources. Moreover, NOAA experts begin to collect data to assess natural resource damages during response operations.

Prevention and Preparedness

Regarding oil spill prevention and preparedness duties, jurisdiction is determined by the potential sources (e.g., vessels, facilities, pipelines) of oil spills. A series of executive orders (EOs), coupled with memoranda of understanding (MOU), have established the various agency responsibilities.[84] **Table 1** identifies the agencies responsible for implementing prevention and preparedness regulations for the potential sources of oil spills.

Prevention responsibilities include, among other things, assessing whether facilities or vessels have the necessary equipment in place. As discussed above, vessels may be required to have double hulls; facilities may need secondary containment.

Preparedness duties involve oversight tasks, such as evaluating facility and vessel response plans. Preparedness responsibilities also include developing and maintaining contingency plans at various levels: area, regional, and national. Personnel training is vital component of sustaining readiness. NOAA oil spill experts help train responders in government service and private business.

In addition, OPA requires agencies to conduct internal examinations to test preparedness.[85] As part of this requirement, the USCG conducts Spills of National Significance (SONS) exercises to analyze the Coast Guard's ability to respond to a major oil spill.

Table 1. Federal Agency Jurisdiction for Oil Spill Prevention and Preparedness Duties, by Source

Potential Source of Oil Spill	Responsible Agency
Vessels	USCG
Onshore, non-transportation facilities	EPA
Onshore, transportation facilities	USCG and Department of Transportation (DOT)
Deepwater ports[a]	USCG and DOT
Offshore facilities (oil/gas extraction)	Minerals Management Service (MMS) within the Department of Interior
Offshore pipelines directly associated with oil extraction activities (i.e., "production lines")[b]	MMS
Offshore pipelines *not* directly associa-ted with oil extraction active-ties (i.e., "transmission lines")	Office of Pipeline Safety (OP S) within the DOT
Inland pipelines	OPS

a. There is only one deepwater port in U.S. coastal waters: the Louisiana Offshore Oil Port (LOOP).

b. For further discussion on federal pipeline jurisdiction, see National Research Council, *Improving the Safety of Marine Pipelines*, National Academies of Science, 1994, pp. 86-89.

Federal Funding for the Oil Spill Liability Trust Fund

In recent years, the level of funding for the trust fund has created some concern, highlighting a central policy debate: how should policymakers allocate the costs associated with a major, accidental oil spill? What share of costs should be borne by the responsible party (e.g., oil vessel owner/ operators), the oil industry, and the general treasury?

Background

Prior to two separate actions by the 109th Congress, the trust fund was particularly vulnerable. In 2005, Congress reinstated the 5-cent-per-barrel tax on oil, thus providing a dedicated source of revenue for the trust fund.[86] In 2006, Congress raised the vessel liability limits, thus requiring the responsible party to pay a greater proportion of the oil spill costs.[87] Before these changes, fund managers projected the fund would be completely depleted by FY2009.[88] With the 2005 legislation in effect, the most recent projection (June 2007) indicates that the fund will reach $1 billion by FY2008 (**Figure 3**). However, this projection does not account for possible claims arising from recent spills, including the 2007 spill into San Francisco Bay, the 2008 spill into the Mississippi River, or oil spills associated with Hurricane Katrina. Although the Katrina-related federal response costs were covered under the Disaster Relief Fund,[89] private parties can seek reimbursement from the trust fund for cleanup or other costs (e.g., damage to property) that exceed their liability limits. The possible number and value of these potential claims remain uncertain.[90]

Regardless of the possible Katrina-related claims, and even considering the substantial enhancements made by the 109th Congress, the trust fund remains vulnerable to a major spill. After conducting two recent Spills of National Significance (SONS) exercises (in 2002 and 2004), the Coast Guard observed that a major spill can exceed the available resources in the trust fund.[91]

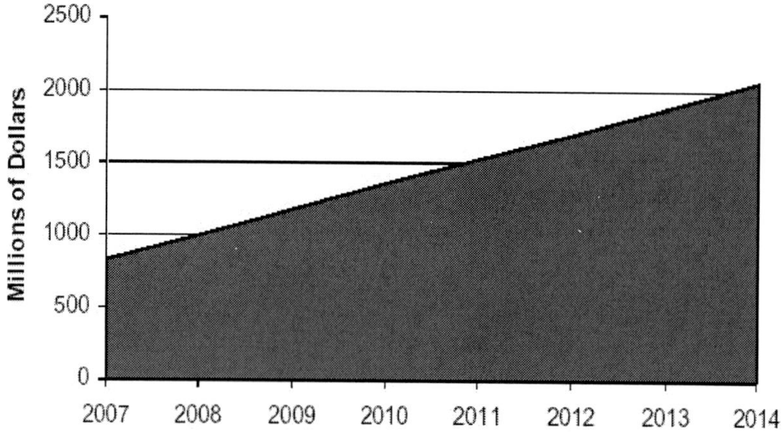

Source: Prepared by CRS with data from National Pollution Funds Center, Oil Spill Liability Trust Fund (OSLTF), Annual Report FY 2002 — FY 2006, at [http://www.uscg.mil/npfc/docs/PDFs/ OSLTF_Report_FY02-FY06.pdf].

Note: The projections assume there will not be any claims associated with spills caused by Hurricane Katrina, the 2007 spill in San Francisco Bay, or the 2008 spill into the Mississippi River.

Figure 3. Projected Annual Balances for the Oil Spill Liability Trust Fund: FY2007 - FY2014

As a reference point, the *Exxon Valdez* spill tallied approximately $2 billion in cleanup costs and $1 billion in natural resource damages (not including third-party claims). If an accidental oil spill from a vessel matching the size of the *Exxon Valdez* (95,000 gross tons)[92] were to occur, the vessel liability (under the new limits) would be capped at either $285 million (single-hull) or $181 million (double hull). Cleanup costs and damages in excess of the liability limits could be covered by the trust fund.[93] OPA Section 9001[94] established per-incident expenditure caps: no more than $1 billion (or the maximum amount available in the fund) for all eligible costs, and no more than $500 million for natural resource damages. Thus, a major spill, particularly one in a sensitive environment, could threaten the viability of the fund, possibly depleting it entirely.

In addition, the above scenario assumes that the owner/operator can take advantage of the liability limits provided by OPA (e.g., obeyed relevant regulations and did not act with "gross negligence or willful misconduct"). If liability is not capped, the owner/operator could be liable for all costs. However, in the case of a significant spill (akin to the *Exxon Valdez*), some owner/operator companies may go bankrupt before paying for the entire cost of the spill. Such a scenario may also threaten the trust fund.

Potential Options

The following options are among several that Congress might consider to address these concerns:

- First, Congress could maintain the status quo, in which the costs are shared between responsible parties, the oil industry, and general taxpayers. Responsible parties — i.e., owners/operators of vessels and facilities — are liable up to their liability caps (if applicable); the trust fund, which is funded through the tax on the oil industry, covers costs above liability limits; general treasury funds would cover costs if the trust fund is depleted.
- Second, Congress could further increase the liability limits for vessels, so that the responsible party would be required to pay a greater portion of the total spill cost before accessing trust fund dollars.
- Third, Congress could increase the 5-cent-per-barrel oil tax to more quickly raise the fund's balance closer to its ceiling of $2.7 billion. A fully funded OSLTF would be more capable of absorbing the costs in excess of a spiller's liability limit.
- Fourth, policymakers may consider expanding liable parties to include the owner of the oil being transported. This option was considered during the development of OPA.[95] The Coast Guard Authorization

Act for Fiscal Year 2008 (S. 1892) would, among other things, alter the liability structure of OPA to make owners of oil liable for spills under certain conditions.

Considerations for Policymakers

The above options spotlight a central policy debate: how should policymakers allocate the costs associated with a major, accidental oil spill?

The first option would maintain the status quo, which represents a cost-sharing structure. Currently, responsible parties and the oil industry bear the costs for all but the most significant spills; general treasury funds are needed if the fund is depleted. Some groups may contend this conflicts with the "polluter pays" principle, arguing that general treasury funds should never have to pay for oil spill cleanup or damages. Some combination of options two, three, and four could reduce the possibility of this occurrence.

The second option would help preserve the trust fund by increasing the liability limit of the spiller. Setting liability limits is a challenge because lawmakers must apportion the risk of a costly, accidental spill fairly between the responsible party and the trust fund. The shipping industry would likely argue that a further increase would be disruptive due to the cost of maintaining additional financial assurance. Considering that the limits were increased substantially in July 2006, this option might be the least feasible.

The third option would increase the cost share borne by the oil industry overall. Some would argue that this follows the polluter pays principle. One view holds that the relevant industry — oil transportation and the oil industry in general — should bear the environmental costs that may develop during the course of business. A different interpretation of the "polluter pays" principle finds that only the responsible party, not industry overall, should pay. One with such a view might conclude that the third option — increasing the per barrel tax — is overreaching and unfair, because it increases the costs borne by the entire industry, instead of on the actual responsible party.[96]

There are likely many economic policy arguments both for and against an additional per-barrel tax on oil. For instance, some groups may claim that an additional oil barrel tax will be reflected in higher energy costs (gasoline, home heating oil) for consumers, which could disproportionately affect lower-income households. (In general, these types of tax policy debates are beyond the scope of this report.) However, policymakers might consider whether any negative consequences from an additional tax on industry would outweigh the potential benefits of allocating the costs to the oil industry.

The fourth option, if implemented, would potentially subject oil companies to liability from spills of their own oil. In line with one view of the polluter pays principle, some may argue that oil owners should be at least partially responsible for the various costs associated with spills of their own oil.[97] Oil industry stakeholders would likely argue that this option is unfair, because, for example, they would have limited control (besides selecting a carrier) over the transportation of their cargo. Moreover, the oil industry would likely point out that they are already supporting oil spill cleanup generally, through the 5-cent-per-barrel tax. Thus, they would likely argue that attaching direct liability to cargo owners would be an inequitable distribution of costs.

Policymakers may find the fourth option attractive, because of the financial strength — or perceived financial strength — of the oil companies.[98] If the United States faces a multi-billion dollar oil spill (and the vessel owner/operators are unable to pay), some may argue that the oil owners should absorb the costs (above the limit of the trust fund) rather than funds from the general treasury. However, this particular liability issue is complicated: further legal and/or policy issues associated with this liability change are beyond the scope of this report.

State Laws

As mentioned above, multiple states had oil spill liability laws before the passage of OPA in 1990. During the 15 years prior to OPA's passage, the issue of whether or not to preempt state liability laws was perhaps the primary obstacle to enacting unified oil spill legislation. Proponents of preemption argued that differing state laws — particularly the various levels of liability — frustrate the shipping industry and are contrary to the goal of comprehensive federal legislation. Preemption opponents maintained that states should be allowed (as with most other federal environmental statutes) to set stiffer standards regarding liability, compensation, and cleanup.[99] In the aftermath of the *Exxon Valdez* spill, the scales tipped to the side of anti-preemption. According to OPA Section 1018 (referred to as a "savings clause"), the act will not preempt any state from imposing additional liability or requirements with respect to the discharge of oil or related response activity (e.g., cleanup standards). A 2003 study identified 16 states that impose unlimited liability for oil spills.[100]

There was some concern that the language of OPA's savings clause would allow states to regulate matters typically reserved for the federal government, such as oil tanker construction. To address this issue, the conference report stated that the savings clause would not disturb a 1978 Supreme Court decision that dealt with the intersection of federal and state authority to regulate the shipping industry.[101] In that case, the Court invalidated a Washington State law that had attempted to govern oil tanker design, size, and movement in Puget Sound.[102]

Regardless of the clarification in the conference report, the line between federal and state jurisdiction continues to be tested. In 2000, the Supreme Court struck down a Washington State rule calling for various personnel requirements, such as training, on oil tankers.[103] Similarly, in July 2006, a federal district court in Massachusetts ruled against a state law that would govern logistics, such as tanker design, personnel qualifications, and navigation.[104]

THREAT OF FUTURE OIL SPILLS IN U.S. COASTAL WATERS

This section examines the threat of future oil spills in coastal waters by assessing the possibilities for such spills, including spill location and oil type (crude versus refined product) and the nation's level of preparedness (or readiness) for responding to a major spill.

Possibilities for Future Oil Spills

Oil is expected to remain the dominant source of energy in the United States for at least the next several decades.[105] Although earlier projections indicated a steady rise in both oil imports and consumption,[106] the most recent estimate (June 2008) from the Energy Information Administration forecasts that oil imports will decline slightly (by about 8%) and then level off in coming years; consumption continues to increase, but at a slower rate (**Figure 4**).[107] *Assuming the projected decline in oil imports actually occurs*, the potential threat of future oil spills from tankers and tank barges may decline to some degree. However, the United States will continue to import a substantial percentage (projected to be between 50% and 60%) of the oil it consumes.[108]

Nonetheless, a potential for spills does not necessarily correspond with a higher spill frequency. U.S. oil consumption has been steadily rising for at least two decades,[109] during which time the number of oil spills in coastal waters has declined (**Figure 1**). Between 1995 and 2004, the number of annual spills declined by approximately 50% (**Figure 5**). Between 2002 and 2004 (the most recent three-year span for which data are available), the annual number of spills in coastal waters from all combined sources remained consistent, averaging 263 incidents per year.[110] This three-year span also shows that while the number of annual spills may remain constant, the annual spill volume can fluctuate. The 2004 spill volume was almost four times greater than the volume spilled in 2003, yet the number of spill incidents was nearly identical. As mentioned above, early data[111] indicate that spill volume in 2005 was substantially higher than recent years, due to spills associated with Hurricanes Katrina and Rita.

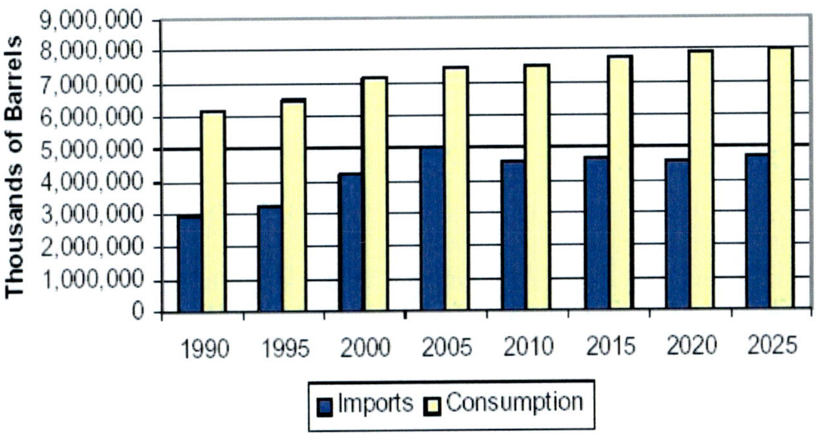

Source: Prepared by CRS with data from the following: actual import data (1990-2005) from Energy Information Administration (EIA) online statistics, at [http://www.eia.doe.gov], and projected import data (2010-2025) and all consumption data from EIA *Annual Energy Outlook 2008* (table 11).

Figure 4. U.S. Petroleum Consumption and Imports: 1990-2005 (Actual), 2010-2025 (Projected)

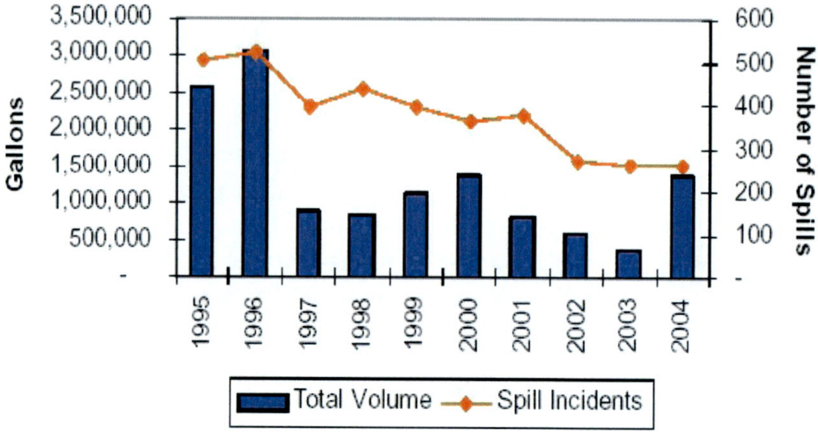

Source: Prepared by CRS with data from the USCG Oil Spill Compendium.

Figure 5. Volume and Number of Oil Spills for Incidents Above 100 Gallons in U.S. Coastal Waters, 1995-2004

Other variables may play a role in the oil spill equation. Oil vessel upgrades (i.e., double hulls) are still ongoing: two important deadlines, 2010 and 2015,[112] will yield a higher percentage of double-hulled vessels in U.S. waters. These improvements may help keep the number of spills low.

Pipeline transportation may increase if imported oil decreases (assuming oil consumption continues to increase). Approximately 170,000 miles of oil pipeline in the United States carry more than 75% of the nation's crude oil and around 60% of its refined petroleum products.[113] The nation's oil pipeline infrastructure is old, and in some (perhaps many) areas, pipelines are operating well beyond their intended service life.[114] As this infrastructure is increasing utilized, the threat of oil spills may increase. However, pipeline safety and security issues have generated interest in recent years, and Congress has addressed some concerns in legislation. For more information on these developments, see CRS Report RL33347, *Pipeline Safety and Security: Federal Programs*, by Paul Parfomak.

U.S. Oil Imports and Possible Spills

The United States will obtain its oil from a combination of domestic and foreign sources. The ratio of this mix and the method of its delivery could play a role in the number and volume of oil spills in the coming years. The following sections discuss this delivery system and its implications for oil spills in greater detail.

Only Canada delivers oil by pipeline to the United States. Although Canada is the leading single source of U.S. oil imports,[115] the vast majority of U.S. oil imports arrive via vessel. (See **Figure 6.)**

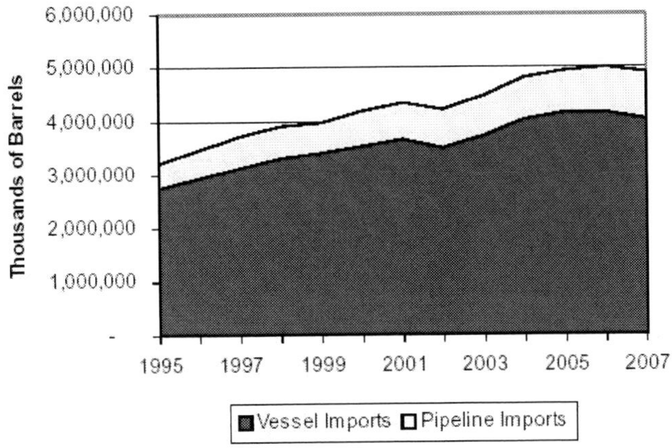

Source: Prepared by CRS with data from EIA online statistics (U.S. Imports by Country of Origin), at [http://www.eia.doe.gov].

Note: Pipeline imports represent imports from Canada; vessel imports are from all other nations.

Figure 6. U.S. Petroleum Imports by Mode of Transportation (1995 - 2007)

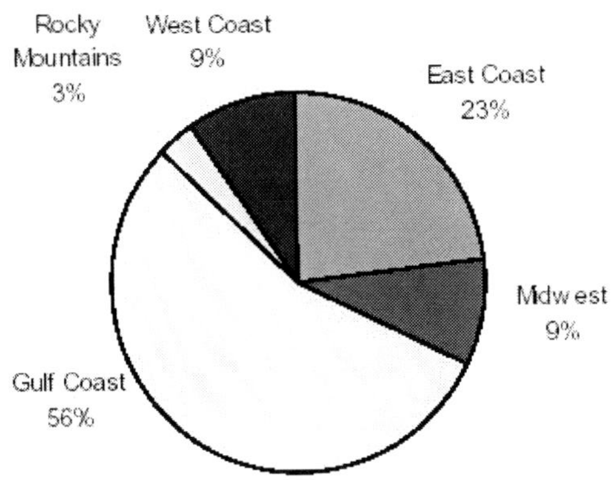

Source: Prepared by CRS with data from EIA online statistics (Imports by Area of Entry), available at [http://www.eia.doe.gov].

Note: The above reflects an average of the years 2000-2007.

Figure 7. Average Annual Distribution of U.S. Oil Imports by Geographic Region

Most of the U.S. oil imports (55%) arrive via the Gulf Coast (**Figure 7**).[116] Of the oil spills within the Coast Guard's jurisdiction (i.e., marine and coastal areas), approximately 50% of the incidents and the volume spilled have occurred in the Gulf of Mexico and its shoreline states.[117] The Gulf will likely remain an area of special attention, because the number of vessels arriving with oil will increase in coming years.

Domestic Oil Transportation and Possible Spills

Although U.S. imports and consumption increased between 1990 and 2005, *total* domestic transportation of crude oil and petroleum products (measured in billions of ton miles)[118] declined by 16% during this time (**Figure 8**). At first glance, this information appears incongruous; however, further analysis reveals that the overall decline is attributable to a 50% decrease in Alaskan oil production that has occurred between 1990 and 2005.[119] Alaskan oil travels long distances via pipeline (across Alaska) and vessel (to the West Coast of the United States) and has historically accounted for a significant percentage of oil in domestic transportation. Therefore, the production of Alaskan oil and its distance traveled will have a strong influence on the overall measure of domestic transportation of oil.

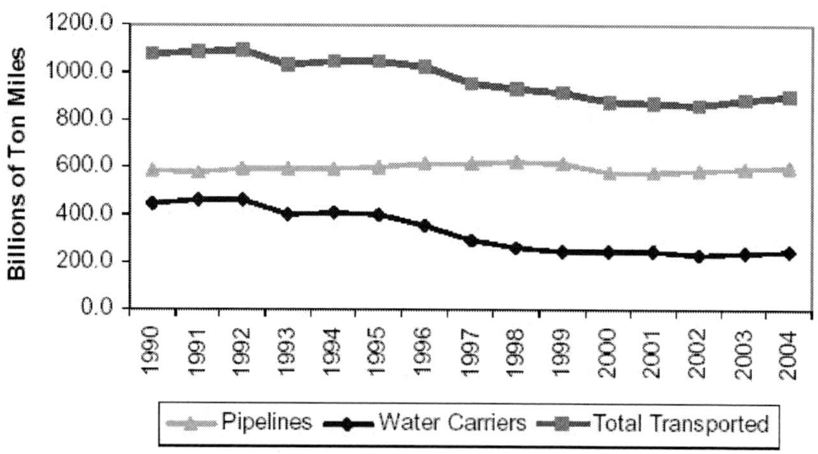

Source: Prepared by CRS with data from Association of Oil Pipelines, *Shifts in Petroleum Transportation*, June 2006, available at [http://www.aopl.org].

Figure 8. Domestic Transportation of Crude Oil and Petroleum Products by Mode, 1990-2004

In recent years, total domestic transportation has shown a modest increase (**Figure 8**). Although the total increase is relatively small,[120] transportation for the continental United States is likely increasing more substantially, considering the context of a continuous decline from Alaskan oil production.[121] As mentioned previously, an increase in oil transportation will not necessarily yield more oil spills, but it may increase the possibilities for spills to occur.

Level of Preparedness

Although many consider the United States' spill record over the last 15 years, particularly for major spills, to have been favorable, the absence of a major spill raises at least one issue. Because the nation has not been forced to respond to a major oil spill for such a long period, some have voiced concern that the nation might have lost the expertise and institutional knowledge necessary for quick and effective response action. The USCG found evidence to support this concern after the 2004 Spill of National Significance (SONS) exercise.[122] The After Action Report concluded:

> Oil spill response personnel did not appear to have even a basic knowledge of the equipment required to support salvage or spill cleanup operations.... There was a shortage of personnel with experience to fill key positions. Many middle-level spill management staff had never worked a large spill and some had never been involved in an exercise. As a result, some issues and complex processes unique to spill response were not effectively addressed.[123]

Congress might consider oversight to monitor this situation as the USCG conducts periodic SONS exercises. The most recent national exercise occurred in June 2007; the "After Action" report was scheduled for release in October 2007, but at the date of this report, the After Action report had not been published.[124]

RECENT OIL SPILL INCIDENTS

November 2007 Oil Spill in San Francisco Bay

On November 7, 2007, a 900-foot container ship — the *Cosco Busan*, flagged as a Hong Kong vessel[125] — struck the San Francisco Bay Bridge, spilling approximately 53,000 gallons of oil into the bay.[126] The spill received considerable attention in the national media, and interested parties have criticized the Coast Guard's response to the incident. The spill led to multiple hearings in the Senate[127] and House,[128] addressing either specifics of the spill or federal law and policies relating to oil spill response, prevention, and preparedness.

During the hearings, Members discussed several issues that may generate further interest and debate concerning oil spill response and prevention. Perhaps the primary issue was the threat of oil spills from non-tank vessels and how the threat could be addressed. Members have introduced several proposals (discussed below) along these lines.

July 2008 Oil Spill into Mississippi River

On July 23, 2008, a tank barge (DM932) being pushed by a tug boat (*Mel Oliver*) collided with a chemical tanker (MV *Tintomara)*. The tank barge was cut into two pieces by the collision. An estimated 380,000 gallons of heavy fuel oil (bunker fuel) were spilled into the Mississippi river,[129] approximately 100 miles north of New Orleans. Officials closed a 100-mile section of the river, and did not fully reopen the section to all vessels for six days. Commerce on the river and at the Port of New Orleans was affected. The incident is still under investigation.

RECENT LEGISLATIVE ACTIVITY

From 1990 to 2005, oil spill issues received relatively little congressional attention. A major spill (more than 1 million gallons) has not had an impact on the nation's shores and sensitive natural resources since 1990. However, the 109th Congress enacted two important provisions affecting cleanup funding and liability.

- P.L. 109-58, H.R. 6, The Energy Policy Act of 2005 (signed August 8, 2005): Congress reinstated the 5-cent-per-barrel tax on domestic and imported oil that had expired on December 31, 1994. As with the original tax, the tax revenue provides direct funding to the Oil Spill Liability Trust Fund (OSLTF) but would be suspended for any period in which the fund's quarterly balance exceeded $2.7 billion (the original suspension trigger was $1 billion).
- P.L. 109-241, H.R. 889, The Coast Guard and Maritime Transportation Act of 2006 (signed July 12, 2006): Congress increased vessel liability limits to $1,900/gross ton for double-hulled vessels and $3,000/gross ton for single-hulled vessels.

In addition, the 109th Congress passed the Pipeline Safety Improvement Act of 2006 (P.L. 109-468), which calls for improved pipeline safety and security practices, and reauthorizes the federal Office of Pipeline Safety.[130] Members introduced other bills that addressed or included provisions concerning oil pollution issues, but none of the bills were considered on the House floor.

In the 110th Congress, Members have introduced several bills that would address oil spill issues. One bill has been enacted; one bill passed the House; several have been reported by various committees:

- P.L. 110-161: The Consolidated Appropriations Act, 2008, was signed by the President December 26, 2007. In the Joint Explanatory Statement of this act, Congress directs the Inspector General of the Department of Homeland Security to prepare an investigative report regarding the November 2007 oil spill in San Francisco Bay.[131] The report was submitted to Congress April 9, 2008.
- S. 184: Senator Inouye introduced this bill January 4, 2007. It would require federal plans for critical pipeline security and incident recovery, and would mandate pipeline security inspections and enforcement. The Senate Committee on Commerce, Science, and Transportation reported the legislation (S.Rept. 110-29) February 15, 2007.[132]
- H.R. 2830: Representative Oberstar introduced this bill June 22, 2007. This legislation would, among other things, decrease the time period from six to three years for submitting claims to the OSLTF. The House Committee on Transportation and Infrastructure reported the bill (H.Rept. 110-338, Part I) September 20, 2007; the Committee on Homeland Security reported the bill October 1, 2007 (H.Rept. 110-

338, Part II); and the Committee on the Judiciary reported the bill October 30, 2007 (H.Rept. 110-338, Part III). This bill passed the House April 24, 2008.

- S. 1566: Senator Inhofe introduced this bill June 7, 2007. It was reported by the Senate Committee on Environment and Public Works June 27, 2008 (S.Rept. 110-403). It would require the U.S. Comptroller General to prepare an annual audit regarding the funds dispersed by the National Pollution Funds Center. Each federal agency that received such funding would also provide an annual report documenting the funding distribution.
- S. 1620: Senator Cantwell introduced this bill June 14, 2007. This legislation would strengthen the current prevention, response, and spill tracking duties of the USCG. The bill would also require certain tank vessel owners to demonstrate financial responsibility to cover potential oil spill costs.
- S. 1892: On July 26, 2007, Senator Cantwell introduced the Coast Guard Authorization Act for Fiscal Year 2008. It was reported by the Senate Committee on Commerce, Science, and Transportation February 5, 2008 (S.Rept. 110-261). Among other things, this bill would direct the Coast Guard to develop regulations covering oil transfers to or from vessels; prepare a report identifying the role of human error in oil spills; and develop an oil spill prevention and education program for small vessels. In addition, the bill would alter the liability structure by (1) making owners (in addition to vessel owners/operators) of transported oil liable for spills from single-hulled vessels after 2010; and (2) broadening the applicability for financial responsibility to tank vessels over 100 gross tons.[133]
- S. 2429: Senator Boxer introduced this bill December 6, 2007, in the wake of the *Cosco Busan* oil spill. This legislation would raise the liability limit for cargo vessels to equal the liability limit of tank vessels.
- S. 2430: Also related to the *Cosco Busan* spill, Senator Boxer introduced this legislation December 6, 2007. The bill would authorize Vessel Traffic Service (operated by the Coast Guard) to direct a vessel pilot to modify the speed or direction of a vessel in an emergency or in hazardous conditions.
- S. 2642: Senator Klobucher introduced the American Renewable Energy Act of 2008 February 14, 2008. This bill would increase the tax rate supporting the OSLTF and extend the tax through 2017.

- S. 2699: Senator Lautenberg introduced the Oil Spill Prevention Act of 2008 March 4, 2008; it was ordered to be reported from the Senate Committee on Commerce, Science, and Transportation May 15, 2008. This bill would, among other things, mandate that new U.S. vessels carrying fuel for propulsion (as opposed to cargo) comply with an amendment — Regulation 1 2A, which went into force August 2007 — to MARPOL. This amendment requires that fuel tanks be protectively located.

APPENDIX. ADDITIONAL STATISTICAL INFORMATION REGARDING OIL SPILLS

Sources of Oil Inputs to U.S. Coastal Waters

Oil enters the coastal waters of the United States from a wide variety of sources. These sources comprise four major categories: natural seeps, oil consumption, oil transportation, and oil extraction.[134] The majority of oil inputs are from natural seeps — geologic openings on the ocean floor. (See **Figure A-1**.) Well-known natural seeps are found in the Gulf of Mexico and off the coast of southern California, regions with extensive oil exploration and production. Although the seeps release large volumes of oil each year,[135] the surrounding ecosystem can adapt, and even thrive, because the rate of release is relatively slow.[136]

Releases associated with petroleum consumption activities account for the vast majority of oil introduced to the coastal environment through human behavior. Rivers and man-made conveyances, such as storm-water drains, receive oil from numerous non-point sources (e.g., urban runoff) and carry the oil to coastal waters. These sources are frequent and widespread, but their rates of release are relatively slow and occur over a long period of time. Moreover, both the quantitative value and the environmental fate of these sources are poorly understood.[137]

Oil transportation and oil extraction represent, on an annual average basis, a minor input to coastal waters. (See **Figure A-1**.) Unlike the other categories, which generally release oil at a slow rate over a wider geographic area, transportation and extraction releases can occur as major spills. Statistics for oil vessel transportation are provided in **Figure 2** at the beginning of this report. Spills from oil extraction operations are discussed later in this Appendix.

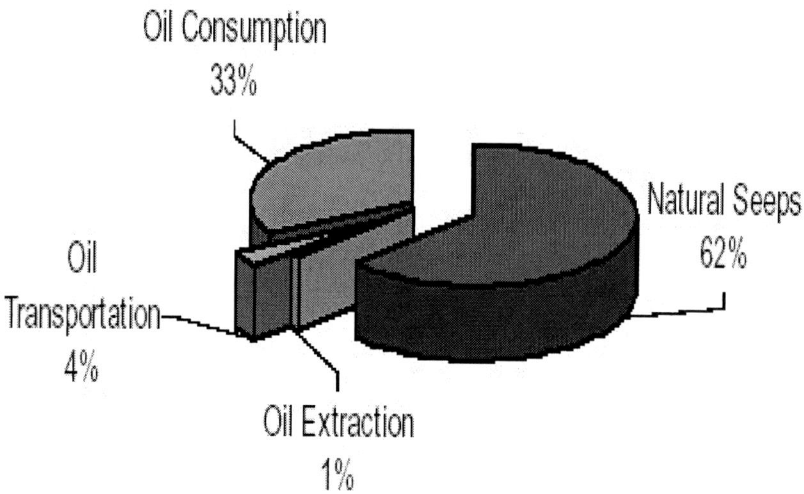

Source: Prepared by the Congressional Research Service (CRS) with data from the National Research Council (NRC) of the National Academies of Science, 2003, *Oil in the Sea III: Inputs, Fates, and Effects*, p. 69.

Figure A-1. Percentage Contribution of Oil Inputs into North American Coastal aters, by Major Source Categories (based on average annual releases, 1990-1999)

Spills from Facilities and Pipelines

The volume of oil spills from facilities and pipelines has declined since the 1980s (see **Figure A-2**). Over the most recent five-year span of data (2000-2004), pipeline spill volume was especially low (compared with previous years), averaging only about 12,000 gallons of spilled oil per year; during the first five-year span (1980-1984), the average annual pipeline spill volume was 2.5 million gallons.

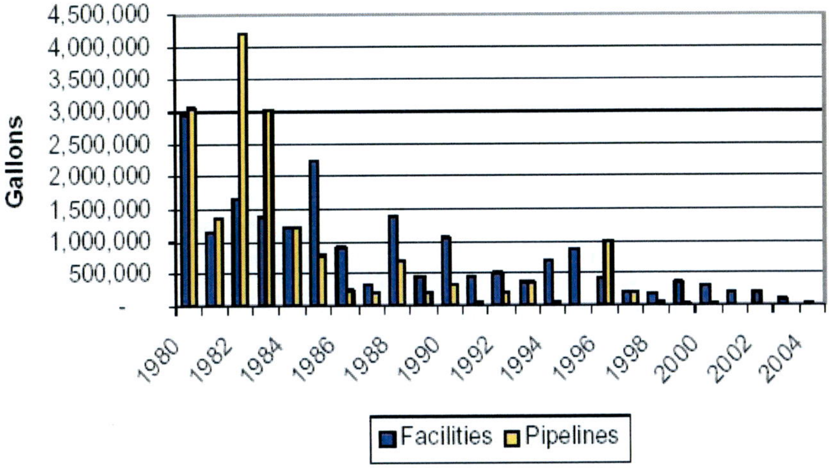

Source: Prepared by CRS with data from USCG Oil Spill Compendium.

Note: The above USCG data includes incidents from land-based facilities and pipelines, as well as oil industry facilities and pipelines in state (nearshore) and federal (offshore) waters.

Figure A-2. Volume of Oil Spills into U.S. Coastal Waters from Facilities and Pipelines, 1980-2004

The number of spills from pipelines in U.S. waters began to decline in the mid-1980s (**Figure A-3**). According to the Coast Guard data, there were only two spills between 2002 and 2004, each spill releasing approximately 15,000 gallons of oil. This low number of pipeline spills may be misleading, because some incidents might not be captured by the Coast Guard's database. In a disclaimer on it website, the Coast Guard points out that its Oil Spill Compendium includes spills "investigated" by the Coast Guard.[138] Some spills in coastal waters may be under another agency's jurisdiction, such as the Office of Pipeline Safety or Minerals Management Service. Although the Coast Guard still includes some of these spills in the database, others are not included.

Spills from facilities have not followed a pattern of steady decline since 1980, but have fluctuated over the last two decades. However, in recent years the number of annual spills has been relatively stable.

Spills from Oil Exploration and Production Operations

As depicted in **Figure A-1**, oil extraction activities contribute approximately 1% of the total oil input to North American waters. The vast majority (95%) of this (1%) oil extraction input comes from operational discharges, which are regulated by a Clean Water Act permit system. Thus, oil extraction spills represent, on an annual basis, only a relatively minor (only 0.05%) component of the total input to North American waters.

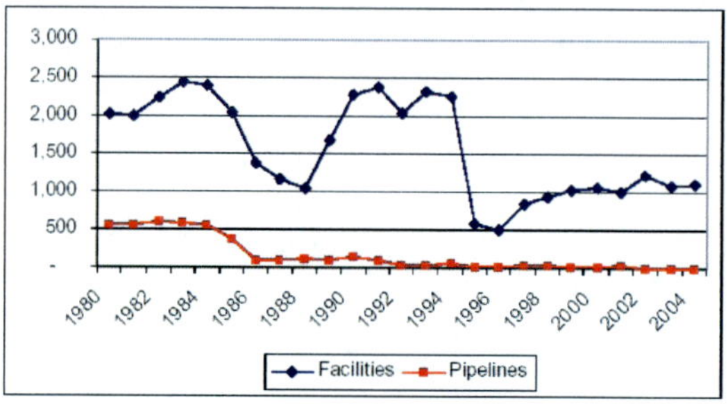

Source: Prepared by CRS with data from the USCG Oil Spill Compendium.

Figure A-3. Annual Number of Spills to U.S. Waters from Facilities and Pipelines, 1980-2004

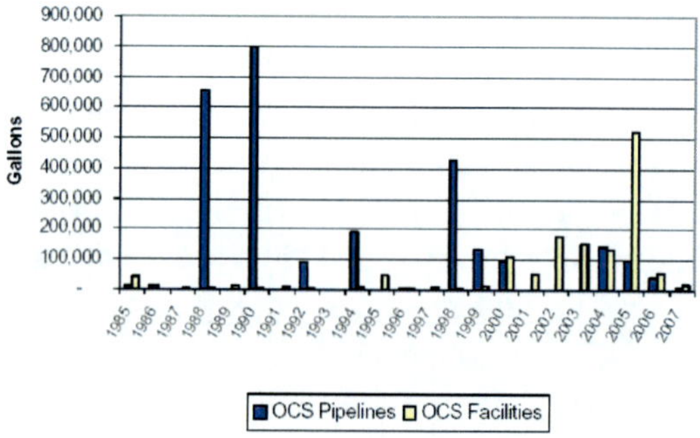

Source: Prepared by CRS with data from the Minerals Management Service (MMS) spill database, at [http://www.mms.gov/incidents/pollution

Note: The MMS database includes spills from chemicals (e.g., methanol), but the above only includes spills of crude oil and petroleum products.

Figure A-4. Annual Oil Spill Volume for Spills Greater than 50 Gallons from Oil Exploration and Extraction Activities in Federal Waters on the Outer Continental Shelf (1985-2007)

However, oil well blowouts from offshore oil extraction operations have historically been a source of major oil spills. The largest accidental oil spill in world history — the IXTOC I, estimated at 140 million gallons — was due to an oil well blowout in Mexican Gulf Coast waters in 1979.[139] A 1969 well blowout off the coast of Santa Barbara released approximately 4 million gallons into the environment and has been credited with catalyzing some of the landmark environmental legislation of the 1970s.

The spill record for offshore platforms in federal waters[140] was noteworthy during the 1980s and 1990s. According to the Minerals Management Service (MMS)[141] oil spill database, there were no oil spills over 1,000 barrels[142] from federally regulated offshore facilities between 1981 and 2001 (**Figure A-4**). This effort was achieved in the context of increasing oil production on the outer continental shelf (OCS).[143]

Since 2002, there have been six offshore facility spills in federal waters over 1,000 barrels, the largest estimated at 2,000 barrels (84,000 gallons). Annual spill volumes from offshore facilities in federal waters have increased in recent years (**Figure A-4**). Part of the recent increase in facility spill volume is attributable to a more comprehensive reporting regime. For example, tanks or machines with oil that are lost at sea during hurricanes are now counted as spills.[144]

As **Figure A-4** indicates, there were several years in which offshore pipelines contributed a relatively minor (or zero) spill volume. However, periodic large spills generated the vast majority of the spill volume from offshore operations between 1985 and 2000. For example, in 1988, there was only one pipeline spill over 1,000 barrels, but that one incident released 15,576 barrels into the environment (654,192 gallons).

The vast majority of the offshore spills in federal waters have taken place in the Gulf of Mexico. In fact, there have been only four spills greater than 50 barrels (2,100 gallons) in the Pacific Region since 1985, none occurring after 1996.[145]

Although **Figure A-4** and the above discussion pertain only to federal waters, there are numerous platforms and pipelines in state waters as well. The majority of the oil extraction operations are located in state waters off the coast of Louisiana and Texas. The precise volume and incident frequency in state waters are difficult to determine. The NRC report describes the data as "generally lacking,"[146] but the report estimates that oil spills from operations in state waters account for twice the oil discharges of activities in federal waters.[147] This estimate is noteworthy considering that spills in state waters present a potentially greater threat to the more sensitive shoreline environments.

End Notes

[1] Note that the *Exxon Valdez* spill ranks only 35th for spill volume on the list of international tanker spills since 1967. See International Tanker Owners Pollution Federation Limited, Historical Data, at [http://www.itopf.com/stats.html].

[2] See National Research Council (NRC), *Oil in the Sea III: Inputs, Fates, and Effects*, National Academies of Science (hereinafter "NRC report"), February 2003, p. 11.

[3] In this report, "oil" refers to crude oil and petroleum products, including gasoline and other fuels, unless stated otherwise.

[4] For the purposes of this report, "U.S. coastal waters" is defined broadly to encompass all waters between the shore and the boundary of the U.S. exclusive economic zone (200 nautical miles from shore). Note that in other documents, "coastal" may refer only to state waters, but in this report, the term "coastal waters" includes state and federally regulated waters.

[5] P.L. 96-478, 33 U.S.C. 1901, et seq. These standards and the U.S. law are discussed later in this report.

[6] Exxon has so far paid approximately $3 billion for the spill: $2 billion for cleanup activities and $900 million in a civil settlement for natural resource damages. In June 2006, parties filed for an additional $92 million for damages per a reopener provision in the civil settlement. On June 25, 2008, the Supreme Court ruled on punitive damages, an issue that had been in the court system for more than a decade. The Court awarded damages of $507 million (a 2006 ruling from the U.S. Court of Appeals for the 9th Circuit had set the damages at $2.5 billion — a 50% reduction from the original, 1994 ruling). The case is still in the system, because plaintiffs are seeking interest on the awarded damages, which could amount to an additional $488 million.

[7] CRS Report (out-of-print, available from CRS by request), *Liability Provisions in State Oil Spill Laws: A Brief Summary*, October 1, 1990.

[8] For example, California passed the Lempert-Keene-Seastrand Oil Spill Prevention and Response Act in 1990. More information is available at [http://www.dfg.ca.gov/ospr/about/history.html#].

[9] NRC report, p. 4.

[10] These "sub-lethal" effects can occur at concentrations that are several orders of magnitude lower than concentrations that cause death. NRC report, p. 127.

[11] NRC report, Chapter 5; also multiple conversations with National Oceanic Atmospheric Administration (NOAA) personnel (2008).

[12] NRC report, p. 120.

[13] NRC report, p. 121.

[14] NRC report, p. 134.

[15] The average cleanup cost is three times higher in the United States than in Europe (based on 1997 data and *excluding* the Exxon Valdez costs). See, Etkin, Dagmar, "Estimating Cleanup Costs for Oil Spills," paper presented at the 1999 International Oil Spill Conference, 1999, citing data from the Oil Spill Intelligence Report International Oil Spill Database.

[16] This is primarily due to the fact that a spill of any size (e.g., in a sensitive area) will require that equipment and response experts be sent to the scene. See Etkin, Dagmar, "Estimating Cleanup Costs for Oil Spills," paper presented at the 1999 International Oil Spill Conference, 1999, p. 5.

[17] For more information, see NOAA's Damage Assessment, Remediation, and Restoration Program at [http://www.darrp.noaa.gov/about/index.html].
[18] 33 U.S.C. Section 2706(d).
[19] 61 *Federal Register* 440 (January 5, 1996). See also NOAA, *Injury Assessment Guidance Document for Natural Resource Damage Assessment Under the Oil Pollution Act of 1990* (1996).
[20] See 15 CFR Section 990.30, definition of "value."
[21] 33 U.S.C. Section 2706(f); William D. Brighton, *Natural Resource Damages under the Comprehensive Environmental Response, Compensation, and Liability Act* (2006), U.S. Department of Justice, Environment and Natural Resources Division.
[22] William D. Brighton, *Natural Resource Damages under the Comprehensive Environmental Response, Compensation, and Liability Act* (2006), U.S. Department of Justice, Environment and Natural Resources Division.
[23] The official statutory name is the Federal Water Pollution Control Act, P.L. 92-500, as amended, codified at 33 U.S.C. 1251, et seq.
[24] P.L. 93-627, codified at 33 U.S.C. 1501, et seq.
[25] P.L. 93-153, codified at 43 U.S.C. 1651, et seq.
[26] P.L. 95-372, codified at 43 U.S.C. 1801, et seq.
[27] See EPA "National Contingency Plan Overview" at [http://www.epa.gov/emergencies/content/lawsregs/ncpover.htm].
[28] The NCP is codified at 40 CFR Part 300.
[29] See, for example, U.S. Congress, House Committee on Merchant Marine and Fisheries, Report accompanying H.R. 1465, Oil Pollution Prevention, Removal, Liability, and Compensation Act of 1989, 1989, H.Rept. 10 1-242, Part 2, 10 1st Cong., 1st sess., p. 32.
[30] The two agreements under consideration were the 1984 Protocols to the International Convention on Civil Liability for Oil Pollution Damage and the Protocols to the International Fund for Compensation for Oil Pollution Damages.
[31] A handful of other oil spills followed the *Exxon Valdez* in 1989 and 1990 (e.g.,the *Mega Borg* spilled 5 million gallons of oil in the Gulf of Mexico), further spurring congressional action.
[32] P.L. 101-380, primarily codified at U.S.C. 2701, et seq.
[33] See, for example, Wilkinson, Cynthia et al., "Slick Work: An Analysis of the Oil Pollution Act of 1990," *Journal of Energy, Natural Resources, and Environmental Law*, 12 (1992), p. 190.
[34] See, Grumbles, Benjamin, and Manley, Joan, "The Oil Pollution Act of 1990: Legislation in the Wake of a Crisis," *Natural Resources and Environment*, 10:2 (1995), p. 38.
[35] U.S. Congress, House Committee on Merchant Marine and Fisheries, Report accompanying H.R. 1465, Oil Pollution Prevention, Removal, Liability, and Compensation Act of 1989, 1989, H.Rept. 10 1-242, Part 2, 10 1st Cong., 1st sess., p. 84.
[36] OPA Section 1011.
[37] OPA Section 4202, amending Section 311(j) of the CWA.
[38] OPA Section 4201(b), amending Section 311 (d)(2)(J) of the CWA.
[39] The response plan requirement is applicable only to an onshore facility that, because of its location, could reasonably be expected to cause substantial harm to the environment by discharging into navigable waters, adjoining shorelines, or the exclusive economic zone. CWA Section 31 1(j)(5)(iii).
[40] OPA Section 4202, amending Section 31 1(j)(5)(E) of the CWA.
[41] U.S. Congress, House Committee on Merchant Marine and Fisheries, Report accompanying H.R. 1465, Oil Pollution Prevention, Removal, Liability, and Compensation Act of 1989, 1989, H.Rept. 101-242, Part 2, 101st Cong., 1st sess., p. 87. OPA Section 4202, amending Section 311 (j)(5)(C)(iii) of the CWA.
[42] Amendments Relating to the Oil Pollution Act of 1990, Title VII of Coast Guard and Maritime Transportation Act of 2004 (P.L. 108-293), codified at 33 U.S.C. 1321.

[43] Primarily the shipboard oil pollution emergency plans required by MARPOL 73/78, discussed later in this report.

[44] A study from the National Academy of Sciences reached this conclusion in 1999. See National Research Council, *Double hull Tanker Legislation: An Assessment of the Oil Pollution Act of 1990*, National Academies of Science, 1999, p. 144.

[45] Opponents maintained that if water entered the space between hulls, the ship could become unstable, hindering salvage and possibly capsizing. Wilkinson, Cynthia et al., "Slick Work: An Analysis of the Oil Pollution Act of 1990," *Journal of Energy, Natural Resources, and Environmental Law*, 12 (1992), p. 196.

[46] U.S. Congress, Conference Report accompanying H.R. 1465, Oil Pollution Act of 1990, 1990, Conf.Rept. 101-653, 101st Cong., 2nd sess., pp. 140-141.

[47] OPA Section 4115, amending 46 U.S.C. 3703.

[48] This exception applied to many inland barges.

[49] Lightering is the process of transferring oil from a large vessel to a smaller vessel. This common practice occurs in designated areas that are typically many miles away from shore.

[50] The definition of "facility" is broadly worded and includes pipelines and motor vehicles. OPA Section 1001.

[51] Under the pre-OPA regime (primarily the CWA), a discharge 12 miles beyond shore had to affect the natural resources before liability attached. Under OPA Section 1002, the discharge itself triggers liability. Wilkinson, Cynthia et al., "Slick Work: An Analysis of the Oil Pollution Act of 1990," *Journal of Energy, Natural Resources, and Environmental Law*, 12 (1992), p. 201.

[52] U.S. Congress, House Committee on Merchant Marine and Fisheries, Report accompanying H.R. 1465, Oil Pollution Prevention, Removal, Liability, and Compensation Act of 1989, 1989, H.Rept. 101-242, Part 2, 101st Cong., 1st sess., p. 31.

[53] OPA Section 1002(b)(1).

[54] OPA Section 1002(b)(2).

[55] Section 107(b) of the Comprehensive Environmental Response, Compensation, and Liability Act (CERCLA, commonly known as Superfund), P.L. 96-5 10.

[56] In addition, liability limits are unavailable if the violation of a federal safety, construction, or operating requirement proximately caused the spill. Spillers must also report the incident and cooperate with response officials to take advantage of the liability caps. OPA Section 1004(c).

[57] 37,895 gross tons x $1,200/ton = $45.47 million. Vessel data from United States Coast Guard, *Investigation into the Striking of Submerged Objects by the Tank Vessel Athos I in the Delaware River on November 26, 2004 with a Major Discharge of Oil*, January 2006, p. 4.

[58] Wilkinson, Cynthia et al., "Slick Work: An Analysis of the Oil Pollution Act of 1990," *Journal of Energy, Natural Resources, and Environmental Law*, 12 (1992), p. 188.

[59] U.S. Congress, House Committee on Merchant Marine and Fisheries, Report accompanying H.R. 1465, Oil Pollution Prevention, Removal, Liability, and Compensation Act of 1989, 1989, H.Rept. 10 1-242, Part 2, 10 1st Cong., 1st sess., p. 35.

[60] Omnibus Budget Reconciliation Act of 1986 (P.L. 99-510).

[61] The CWA Section 311(k) revolving fund; the Deepwater Port Liability Fund; the Trans-Alaska Pipeline Liability Fund; and the Offshore Oil Pollution Compensation Fund.

[62] Omnibus Budget Reconciliation Act of 1989 (P.L. 101-239). Other revenue sources for the fund include interest on the fund, cost recovery from the parties responsible for the spills, and any fines or civil penalties collected.

[63] And in accordance with Executive Order 11735 (August 3, 1973), granting EPA the authority to regulate non-transportation-related onshore and offshore facilities.

[64] U.S. EPA, "Oil Pollution Prevention: Non-Transportation Related Onshore and Offshore Facilities," *Federal Register*, vol. 38, no. 237 (December 11, 1973), pp. 34164-34170.

[65] EPA offered several regulatory amendments after the 1973 rulemaking. Following OPA in 1991, the agency proposed substantial changes and clarifications that were not made final

until 2002. The effective date of the 2002 final rule has been extended five times; the current effective date is July 1, 2009. *Federal Register*, vol. 72, no. 94, pp. 27443-27448 (May 16, 2007).

[66] CWA Section 311(j)(1)(C).

[67] See 40 CFR Section 112.1.

[68] See, for example, MMS Press Release from February 2, 2005, at [http://www.mms.gov/ooc/press/2005/press0202.htm].

[69] For a comprehensive list of federal maritime legislation see USCG, *Marine Safety Manual*, Vol. IX (undated), Chapter 1, available at [http://www.uscg.mil/hq/g-m/nmc/pubs/msm/vol9.pdf].

[70] For example, the Rivers and Harbors Act of 1899, as amended (33 U.S.C. 401, et seq.), and the International Regulations for Preventing Collisions at Sea, as amended (33 U.S.C. 1601, et seq.).

[71] P.L. 92-340, 33 U.S.C. 1221, et seq.

[72] P.L. 95-474, codified at 33 U.S.C. 1221-1232 and 46 U.S.C. 3701-3718.

[73] If a treaty is considered "self-executing," domestic legislation implementing the treaty is not necessary. For more details on these issues, see CRS Report RL32528, *International Law and Agreements: Their Effect Upon U.S. Law*, by Michael John Garcia and Arthur Traldi.

[74] For convention texts and other materials, see [http://www.imo.org].

[75] The phrase "entry into force" signifies that the requisite number of nations have ratified the convention or annex, thus making the agreed upon requirements binding for all participating nations. For more discussion of the procedures of international conventions, see the IMO website at [http://www.imo.org].

[76] All vessels of any type over 400 gross tons traveling over international waters must have a SOPEP approved by their flag state. See USCG VRP/SOPEP "FAQs" at [*http://www.uscg. mil/vrp]*.

[77] USCG, 1997, *Marine Safety Manual, Marine Environment Protection, Volume IX*, p. 4-24.

[78] P.L. 96-478, 33 U.S.C. 1901, et seq.

[79] The *Torrey Canyon*, a Liberian-flagged tanker, spilled approximately 35 million gallons of crude oil.

[80] P.L. 93-248 , 33 U.S.C. 1471, et seq.

[81] Most state waters extend 3 nautical miles (1 nautical mile = 6,076 feet, or 1.15 miles) from shore. Louisiana waters extend 3 imperial nautical miles (1 imperial nautical mile = 6,080 feet). Texas and Gulf Coast of Florida waters extend 3 marine leagues (equating to 9 nautical miles). See the MMS, OCS, website ("Definitions and Jurisdictions") at [http://www.mms.gov/incidents/pollution

[82] The terms inland and coastal are defined in the National Contingency Plan (40 CFR Section 300.5). The coastal zone covers all waters subject to the tide, the Great Lakes, and all seaward waters (extending 200 nautical miles beyond shore). The inland zone covers all other U.S. waters. Spills in inland waters can potentially affect coastal waters and ecosystems, particularly if the spill occurs in water systems near the coast. In fact, a fine line may separate specific inland and coastal waters (e.g., consider the nexus between a bay and a river).

[83] For more information see NOAA's Office of Response and Restoration website, at [http://response.restoration.noaa.gov/index.php].

[84] Executive Order (EO) 12777 (October 18, 1991) delegates authorities pursuant to the Oil Pollution Act of 1990. This order was amended by EO 13286 (March 5, 2003), which reorganized duties in response to the creation of the Department of Homeland Security.

[85] As required by OPA Section 4202(a), which amended CWA Section 311(j)(7), codified in 33 U.S.C. 1321(j)(7).

[86] Energy Policy Act of 2005 (P.L. 109-58). Congress also raised the ceiling from $1 billion to $2.7 billion for when the tax would temporarily expire. Under the original tax legislation (Omnibus Budget Reconciliation Act of 1989 (P.L. 101-239), the per-barrel tax would be suspended in any calendar quarter if the fund balance reached $1 billion, restarting again if it dipped below that number. The 2005 act increased this threshold to $2.7 billion.

[87] Coast Guard and Maritime Transportation Act of 2006 (P.L. 109-241).

[88] USCG, *Report on the Oil Spill Liability Trust Fund*, May 2005, p. 11.

[89] The President may issue a major disaster declaration under the Robert T. Stafford Disaster Relief and Emergency Assistance Act (the Stafford Act) and invoke federal authorities and resources, as occurred in response to Hurricane Katrina. For more on this issue, see CRS Report RL33053, *Federal Stafford Act Disaster Assistance: Presidential Declarations, Eligible Activities, and Funding*, by Keith Bea.

[90] The most recent publications from the NPFC, which manages the OSLTF, do not provide an estimate because of this uncertainty. However, a background document that was offered at an April 27, 2006, hearing stated, "the Coast Guard had estimated these costs could exceed $800 million" (House Subcommittee on Coast Guard and Maritime Transportation, Hearing on the Implementation of the Oil Pollution Act). An NPFC official suggests this estimate is possibly overstated (per telephone conversation on September 22, 2006).

[91] See U.S. Department of Homeland Security, U.S. Coast Guard, *California SONS 2004: After Action Report*, pp. 46 and A-7.

[92] The *Exxon Valdez* has been renamed the *Sea River Mediterranean*, and its specifications can be found at the USCG Port State Information Exchange database at [http://cgmix.uscg.mil/PSIX/].

[93] Except for emergency funds available to federal officials, the trust fund distributes monies through a claims process. Persons who have incurred cleanup costs or suffered damages from an oil spill may submit a claim for reimbursement from the trust fund. This includes responsible parties who have reached their respective liability limits. For more details on this process, see the National Pollution Fund Center's website at [http://www.uscg.mil/npfc/].

[94] Codified in 26 U.S.C. 9509.

[95] The House version of the bill — 101st Congress, H.R. 1465 — would have held, under certain conditions, oil owners liable for 50% of the costs and damages.

[96] The Bush Administration maintained an analogous view of the "polluter pays" principle in the context of whether or not to reinstate the Superfund tax on the chemical and oil industry. See CRS Report RL3 1410, *Superfund Taxes or General Revenues: Future Funding Issues for the Superfund Program*, by Jonathan L. Ramseur, Mark Reisch, and James E. McCarthy.

[97] In addition, this change would arguably make OPA liability more in line with the liability structure of the Comprehensive Environmental Response Compensation and Liability Act (CERCLA or Superfund, P.L. 96-510), which governs releases of hazardous substances.

[98] Although record profits for certain oil companies have been highlighted in recent years, it is beyond the scope of this report to assess the extent of profits and financial worth of the oil industry overall, or the percentage of firms that achieved record profits.

[99] In fact, one argument against preemption was that existing requirements under particular state laws would be diminished or negated entirely. See Benjamin Grumbles and Joan Manley, "The Oil Pollution Act of 1990: Legislation in the Wake of a Crisis," *Natural Resources and Environment*, 10:2 (1995), p. 38.

[100] Etkin, Dagmar, 2003, *A Worldwide Review of Marine Oil Spill Fines and Penalties*, at [http://www.environmental-research.com/erc_papers/ERC_paper_10.pdf].

[101] U.S. Congress, Conference Report accompanying H.R. 1465, Oil Pollution Act of 1990, 1990, Conf.Rept. 101-653, 101st Cong., 2nd sess., p. 122.

[102] *Ray v. Atlantic Richfield*, 435 U.S. 151 (1978).

[103] *United States v. Locke*, 529 U.S. 89 (2000).

[104] *United States v. Massachusetts*, 440 F. Supp. 2d 24 (D. Mass. 2006).
[105] Energy Information Administration (EIA), *Annual Energy Outlook 2008* (hereinafter EIA 2008 Report), June 2008, at [http://www.eia.doe.gov/].
[106] See the 2007 version (and earlier editions) of the EIA's *Annual Energy Outlook.*
[107] EIA 2008 Report, Table 11.
[108] EIA 2008 Report, Figure 90.
[109] The increase in U.S. oil consumption may have started even earlier. The EIA historical data only go back to 1981. EIA online petroleum statistics, "Total Crude and Petroleum Products Supplied," (historical data), at [http://www.eia.doe.gov].
[110] See USCG Oil Spill Compendium.
[111] The Coast Guard database contained data through 2004 at the time of this report, but MMS has data related to offshore oil production for 2005.
[112] Barring a few exceptions, single-hull vessels over 5,000 gross tons may not operate in U.S. waters after January 1, 2010. Single-hull vessels below 5,000 gross tons (chiefly barges) and other previously excepted vessels (e.g., lightering vessels or those upgraded with double-sides or bottoms) must have double hulls by January 1, 2015. OPA Section 4115 (codified at 46 U.S.C. 3703a).
[113] Bureau of Transportation Statistics (BTS), "National Transportation Statistics" (2008), at [http://www.bts.gov/publications/national_transportation 1_1 0. html].
[114] See, for example, National Research Council, *Improving the Safety of Marine Pipelines*, National Academies of Science, 1994, p. 45, and Epstein, Paul (editor), *Oil: A Life Cycle Analysis of its Health and Environmental Impacts*, Harvard Medical School's Center for Health and the Global Environment, March 2002, p. 21.
[115] Approximately 18% of total U.S. oil imports came from Canada in 2007. Energy Information Association, Petroleum Statistics, U.S. Imports by Country of Origin, available at [http://www.eia.doe.gov].
[116] Although most crude oil (62%) enters the United States through the Gulf, most of the petroleum products (70%) enter through ports on the east coast. EIA online statistics (Imports by Area of Entry), available at [http://www.eia.doe.gov].
[117] USCG Spill Compendium, Cumulative Data And Graphics For Oil Spills (1973-2004), at [http://www.uscg.mil/hq/g-cp/comrel/factfile/index.htm].
[118] Note that using this unit of measure is different from using volume alone. With this measure, distance traveled is as important a factor as volume.
[119] Alaskan crude oil production yielded 647,309 barrels in 1990 but declined to 315,420 barrels by 2005. Production decline further to 262,434 barrels in 2007 (EIA website, Petroleum, "Crude Oil Production").
[120] The total value increased from 873 billion ton miles in 2000 to 903 billion ton miles in 2004.
[121] The degree of increase for oil transportation within the continental United States is uncertain, because the segregated data are not readily available.
[122] The California SONS 2004 was conducted in April 2004 and was the fourth SONS exercise performed by the USCG. The purpose of the SONS exercises is to ensure and evaluate the readiness of personnel to respond to a major oil spill. The 2004 SONS exercise was the largest to date and the first international SONS (involving Mexico).
[123] U.S. Department of Homeland Security, U.S. Coast Guard, *California SONS 2004: After Action Report*, September 2004, p. 46.
[124] See [https://www.sons-program
[125] Vessel statistics are available in the USCG Port State Information Exchange database at [http://cgmix.uscg.mil/PSIX/].
[126] Department of Homeland Security, Office of Inspector General, *Allision of the M/V Cosco Busan With the San Francisco-Oakland Bay Bridge* (2008).
[127] The Senate Commerce, Science, and Transportation Committee held three hearings: one in the full committee (November 14, 2007); one in the Subcommittee of Oceans, Atmosphere,

Fisheries, and Coast Guard (December 18, 2007); one in the Subcommittee on Surface Transportation and Merchant Marine Infrastructure, Safety, and Security (March 4, 2008).

[128] The House Committee on Transportation and Infrastructure Subcommittee on Coast Guard and Maritime Transportation held hearings on November 19, 2007, and April 10, 2008.

[129] Spill estimate from NOAA website, at [http://response.restoration.noaa.gov].

[130] The President signed the legislation into law on December 29, 2006.

[131] See Congressional Record, December 17, 2007 (Book II), Joint Explanatory Statement, Division E, Title I, House pp. H16080-81.

[132] For further information on pipeline legislation and issues, see CRS Report RL33347, *Pipeline Safety and Security: Federal Programs*, by Paul W. Parfomak.

[133] The current level is 300 tons.

[134] NRC report, pp. 67-88.

[135] The NRC report (p. 69) estimates that natural seeps off North America release 47 million gallons of oil each year (converted from tonnes).

[136] NRC report, p. 2.

[137] The range of uncertainty of land-based runoff is substantial, from a minimum estimate of 5.6 million gallons to 588 million gallons (based on average, annual releases from 1990-1999). NRC report, pp. 69, 87.

[138] See Oil Spill Compendium at [http://www.uscg.mil/hq/g-cp/comrel/factfile/index.htm].

[139] NRC report, p. 33.

[140] Federal waters extend 200 nautical miles from shore but do not include the waters adjacent to shore that are under state control (as discussed earlier in this report).

[141] The MMS, in the U.S. Department of Interior, is the federal agency responsible for managing the oil and gas resources on the outer continental shelf.

[142] 1,000 barrels = 42,000 gallons.

[143] The annual crude oil production in the Gulf of Mexico (the primary source of offshore crude oil) increased by 112% from 1981 to 2001. Energy Information Administration, Crude Oil Production statistics, at [http://www.eia.doe.gov/].

[144] Per August 15, 2006, telephone conversation with MMS official.

[145] MMS Oil Spill database, at [http://www.mms.gov/incidents/pollution

[146] NRC report, p. 193.

[147] NRC report, p. 38.

In: Poisoning and Acidification of the Earth's... ISBN: 978-1-60741-560-2
Editors: Geoffrey E. Mason

Chapter 6

U.S. DISPOSAL OF CHEMICAL WEAPONS IN THE OCEAN: BACKGROUND AND ISSUES FOR CONGRESS

David M. Bearden
Environmental Policy Resources, Science, and Industry Division

SUMMARY

The U.S. Armed Forces disposed of chemical weapons in the ocean from World War I through 1970. At that time, it was thought that the vastness of ocean waters would absorb chemical agents that may leak from these weapons. However, public concerns about human health and environmental risks, and the economic effects of potential damage to marine resources, led to a statutory prohibition on the disposal of chemical weapons in the ocean in 1972. For many years, there was little attention to weapons that had been dumped offshore prior to this prohibition. However, the U.S. Army completed a report in 2001 indicating that the past disposal of chemical weapons in the ocean had been more common and widespread geographically than previously acknowledged. The Army cataloged 74 instances of disposal through 1970, including 32 instances off U.S. shores and 42 instances off foreign shores. The disclosure of these records has renewed public concern about lingering risks from chemical weapons still in the ocean today.

The risk of exposure to chemical weapons dumped in the ocean depends on many factors, such as the extent to which chemical agents may have leaked into seawater and been diluted or degraded over time. Public health advocates have questioned whether contaminated seawater may contribute to certain symptoms among coastal populations, and environmental advocates have questioned whether leaked chemical agents may have affected fish stocks and other marine life. There also has been public concern that chemical weapons could wash ashore or be accidentally retrieved during activities that disturb the seabed, such as dredging and trawl fishing. Although such incidents have occurred domestically and abroad, they are rare relative to the thousands of weapons dumped in the ocean. Assessing the degree of risks is difficult because of a lack of information.

Whether the risks are low or high, how to respond to them is fraught with many challenges. The primary obstacle is locating the weapons in the ocean. The lack of coordinates for most of the disposal sites, and the possibility that ocean currents may have moved weapons beyond these areas, makes finding the weapons difficult at best, if not impracticable in some cases. Enacted in the second session of the 109th Congress, the John Warner National Defense Authorization Act for FY2007 (P.L. 109-364, H.R. 5122) requires further review of historical records to attempt to identify where chemical and conventional weapons were dumped off U.S. shores, research of the effects of these weapons on the ocean environment, and monitoring to determine whether contamination or health or safety risks are present. The funding and implementation of these requirements are potential issues for the 110th Congress.

In the event that the weapons are located, retrieving them from the seabed could be technically challenging and could introduce new risks during retrieval and transport for onshore disposal. Leaving located weapons in place, and warning the public to avoid these areas, may be more feasible and involve fewer immediate risks. However, long-term risks would remain. Responding to potential risks is further complicated by insufficient information to reliably estimate response costs and by the uncertain availability of federal funding to pay for such actions.

INTRODUCTION

Greater awareness of the past disposal of chemical weapons in the ocean has motivated growing concern among the public about potential risks to human health, safety, and the marine environment. The Department of Defense (DOD) reports that the United States ceased the disposal of chemical weapons in the ocean in 1970. Congress later enacted legislation in 1972 that banned the disposal of wastes in the ocean in general, including chemical weapons. Although DOD has indicated that chemical weapons are no longer dumped in the ocean, much is unknown about the potential risks from the past disposal of such weapons still in the ocean today.

A report completed by the U.S. Army in 2001[1] provided more information than previously released on specific areas of the ocean where the U.S. Armed Forces had disposed of chemical weapons.[2] In its report, the Army acknowledged that some of these weapons were damaged or leaking at the time of disposal. In light of this more recent information, public health and environmental advocates, marine conservationists, and the general public have raised questions about the potential risks of chemical weapons in the ocean and have suggested that scientific study is needed to assess these risks.

This report provides a brief history of the disposal of chemical weapons in the ocean by the U.S. Armed Forces, discusses potential risks to human health and the marine environment, reviews findings of relevant scientific studies of risks from the disposal of chemical weapons off the coasts of Europe and Russia, analyzes factors that determine the feasibility of responding to potential risks, identifies possible response authorities in existing federal law, and examines historical review, scientific research, and monitoring requirements in Section 314 of the John Warner National Defense Authorization Act for FY2007 (P.L. 109-364, H.R. 5122).

History of U.S. Disposal of Chemical Weapons in the Ocean

In the late 1960s, DOD first publicly acknowledged that the U.S. Armed Forces had routinely disposed of chemical weapons in the ocean since World War I. In its 2001 report, the Army provided more extensive historical records on the number of instances and areas of the ocean where the U.S. Armed Forces disposed of chemical weapons. The Army catalogued 74 instances of disposal in the ocean, of which 32 were off U.S. shores and 42 were off foreign shores. The first recorded instance was in 1918 at an unknown location in the Atlantic Ocean between the United States and England. The Army's records did not note other instances of ocean disposal until 1941. Therefore, the extent to which ocean disposal may have occurred in between these years is unknown. According to the Army, the last instance of disposal occurred in 1970, approximately 250 miles off the coast of Florida.

Estimating the cumulative quantity of chemical weapons dumped in the ocean, and identifying all types of such weapons, is not possible because of incomplete historical records. The Army's 2001 report indicated that the number of chemical weapons in each instance of disposal ranged widely, from a few weapons to thousands. The Army also indicated that in some instances, conventional explosives and radiological waste were dumped in the ocean along with chemical weapons. The volume of chemical weapons agents also varied widely, from 30 pounds or less to thousands of tons. The types of chemical weapons varied as well, commonly including sulfur mustard[3] and nerve agents.[4] At some sites, the Army does not know the exact substances that were disposed of in the ocean. The reasons for ocean disposal also varied. Some weapons were deemed surplus. Others were damaged and leaking chemical agents, presenting an immediate risk to the military personnel who managed them. Certain weapons were not produced by the United States, but were captured from foreign nations and were disposed of to prevent their use.

The dumping of chemical weapons in the ocean was widespread geographically, including areas off the coast of the continental United States in the Atlantic and Pacific Oceans and the Gulf of Mexico, off the coast of Hawaii, and two instances of disposal in the Mississippi River in Louisiana. Although the Army identified individual instances of disposal by site, the exact coordinates for many of these sites are unknown. Rather, a broad geographic reference to a state or city on the coast and the approximate distance from shore is specified to denote the location. In some instances, only the Atlantic or Pacific Ocean at large is identified. The Army also acknowledged disposal by the U.S. Armed Forces off the coasts of foreign nations in the Atlantic and Pacific Oceans, North Sea, Mediterranean Sea, and Indian Ocean. (See the following table for a list of instances of disposal in the ocean off U.S. shores. A complete list, including disposal off foreign shores, is provided in the Army's 2001 report.)

U.S. Army Records of Past Disposal of Chemical Weapons in the Ocean off the Coasts of the United States through 1970

Date or Time Period	Approximate Location of Disposal and Point of Origin Onshore	Chemical Weapon or Agent
World War II	Atlantic Ocean, off Charleston, South Carolina	105 millimeter (mm) mustard projectiles and M70 115-pound mustard bombs
World War II	Atlantic Ocean, off Norfolk, Virginia	Unknown
August 1944	Pacific Ocean, neither shore-line point of reference nor originating point are specified	36 M47A2 100-pound mustard bombs and approximately 15,000 unspecified bombs
1944	Pacific Ocean, off Pearl Harbor, Oahu, Hawaii	4,220 tons of unspecified toxics and hydrogen cyanide
1944	Pacific Ocean, about five miles off Oahu, Hawaii	approximately 16,000 M47A2 100-pound mustard bombs
1944 (year uncertain)	Mississippi River, originating from New Orleans, Louisiana	More than 20 (number unce-rtain) leaking M70 115-pound mustard bombs
September 14-December 21, 1945	Atlantic Ocean, "Disposal Area Number 1," originating from Edgewood Arsenal, Maryland	1,154 55 gallon drums of arse-nic trichloride 375 tons of diphenylaminechloroarsine (adam-site) smoke candles 75,852 4.2-inch distilled mustard shells 924 M74 10-pound white phos-p-horus cluster bombs approximately 56,000 smoke canisters approximately 23,000 smoke projectiles
October 17-November 2, 1945	Pacific Ocean, off Hawaii, originating from Waianae, Hawaii	20 M79 1000-pound hydrogen cyanide bombs 1,100 M79 1000-pound cyanogen chloride bombs 125 M78 500-pound cyanogen chloride bombs 14,956 M70 114-pound mustard bombs 30,917 4.2-inch mortar mustard shells

Table. (Continued)

Date or Time Period	Approximate Location of Disposal and Point of Origin Onshore	Chemical Weapon or Agent
		1,038 one-ton containers of mustard agent 190 one-ton containers of lewisite
1945	Mississippi River, 3-4 miles south of Braithwaite, Louisiana, as a result of accidental sinking in quicksand in a ship canal	2 unspecified bombs
March 7, 1946	Gulf of Mexico, originating from New Orleans, Louisiana	Unspecified quantity of mustard projectiles
March 10, 1946	Gulf of Mexico, originating from Edgewood Arsenal, Maryland, and loaded for sea disposal at New Orleans, Louisiana	2 leaking mustard bombs
March 21-25, 1946	Atlantic Ocean, "Baker" Site off Charleston, South Carolina, originating from Naval Mine Depot, Virginia	4 "carloads" of mustard projectiles
May 1946	Gulf of Mexico, originating point not specified	3 phosgene bombs (German origin)
July 13, 1946	Gulf of Mexico, 20 miles offshore, originating from Mobile, Alabama	30 500 kg mustard bombs 3 250 kg mustard bombs (German origin)
August 1-October 17, 1946	Atlantic Ocean, "Baker" Site off Charleston, South Carolina	lewisite, mustard, phosgene bombs German mustard and tabun bombs mustard one-ton containers mustard projectiles (quantities unspecified)
June 30- July 15, 1947	Pacific Ocean, 12 miles off Aleutian Islands, originating from Attu and Adak, Alaska	61 containers of mustard agent 887 containers bulk lewisite
December 1520, 1948	Atlantic Ocean, 300 miles off Florida, originating from Gulf Chemical Warfare Depot, Alabama, via Charleston, South Carolina	3,711 containers of lewisite 60 M14 bulk lewisite
January-February 1955	Gulf of Mexico, originating from Mobile, Alabama	"1 or 2 barges" of unspecified toxic munitions (quantity not

		specified)
November 13-14, 1957	Atlantic Ocean, originating from Edgewood Arsenal, Maryland, and loaded for sea disposal at Colts Neck Naval Pier, Earle, New Jersey	48 one-ton containers of lewisite
March 20-27, 1958	Atlantic Ocean, off South Carolina, originating from Pine Bluff Arsenal, Arkansas, and loaded for sea disposal at Sunny Point, North Carolina	1,507 one-ton containers of lewisite 63 one-ton containers of nitrogen mustard
April 19, 1958	Pacific Ocean, 117 miles off San Francisco, California, originating from Navajo Army Depot, Arizona, and Tooele Army Depot, Utah, and loaded for sea disposal at Concord Naval Weapons Station, California	301,000 M70 115-pound mustard bombs 1,479 one-ton containers of lewisite
May 25, 1958	Pacific Ocean, 117 miles off San Francisco, California, originating from Tooele Army Depot, Utah, and loaded for sea disposal at Concord Naval Weapons Station, California	6 M47 100-pound mustard bombs 335 one-ton containers of lewisite 11 one-ton containers of nitrogen mustard 2 mustard projectiles
June 14-15, 1960	Atlantic Ocean, originating from Edgewood Arsenal, Maryland	2 one-ton containers of lewisite 1 lewisite cylinder non-chemical weapons materials (unspecified quantity of radiological waste)
June 18, 1962	Atlantic Ocean, originating from Edgewood Arsenal, Maryland	378 105 mm mustard projectiles 341 155 mm mustard projectiles 1 one-ton container of lewisite 20 drums of cyanide 5,252 white phosphorous munitions
		non-chemical weapons materials (421,757 pounds of radiological waste)
June 15, 1967	Atlantic Ocean, originating from Colts Neck Naval Pier, Earle, New Jersey	4,577 one-ton containers of mustard agent 7,380 M55 sarin rockets in concrete vaults

Table. (Continued)

Date or Time Period	Approximate Location of Disposal and Point of Origin Onshore	Chemical Weapon or Agent
June 19, 1968	Atlantic Ocean, originating from Colts Neck Naval Pier, Earle, New Jersey	38 one-ton containers of sarin and VX 1,460 vaults holding M55 sarin and VX rockets 120 drums of canisters of arsenic and cyanide
August 7, 1968	Atlantic Ocean, originating from Colts Neck Naval Pier, Earle, New Jersey	3,500 one-ton containers contaminated with mustard agent and filled with water non-chemical weapons materials (unspecified quantity and type of conventional explosives)
August 18, 1970	Atlantic Ocean, 250 miles east of Cape Kennedy, Florida, originating from Sunny Point, North Carolina	12,508 M55 sarin rockets in vaults 3 155mm sarin projectiles 1 M23 VX land mine

Source: Prepared by the Congressional Research Service with information from the Department of Defense, U.S. Army Research, Development, and Engineering Command, Aberdeen Proving Ground, Maryland, Corporate Information Office, Historical Research and Response Team, *Off-shore Disposal of Chemical Agents and Weapons Conducted by the United States*, March 29, 2001, 15 pp.

Cessation of U.S. Disposal in the Ocean

In the late 1960s, DOD's acknowledgment ofthe disposal of chemical weapons in the ocean, and heightened public awareness of the ocean disposal of wastes in general, raised concerns about potential risks to human health and the marine environment, and the economic effects of potential damage to marine resources. In light of these concerns, DOD requested that the National Academy of Sciences (NAS) assess the hazards of disposing of surplus chemical weapons, including land and sea disposal. The NAS released a report in 1969 recommending the pursuit of methods to safely destroy or neutralize chemical weapons, rather than bury them intact on land or at sea.[5] These recommendations and continuing public concerns led the United States to cease disposal of chemical weapons in the ocean in 1970 and to explore methods to destroy surplus weapons at military facilities where they could be managed safely.

In 1972, Congress enacted the Ocean Dumping Act[6] to prohibit the disposal of wastes into the ocean waters of the United States, extending to the contiguous zone (24 nautical miles seaward). Consistent with the decision of the executive branch in 1970 to cease the disposal of chemical weapons in the ocean, Congress included provisions in the Ocean DumpingAct that explicitly prohibited the offshore disposal of chemical warfare agents. Although the act granted limited authority for the Environmental Protection Agency (EPA) to issue permits allowing the offshore disposal of certain types of wastes, it specifically excluded chemical warfare agents and other hazardous substances from this permit authority.[7] (See CRS Report RS20028, *Ocean Dumping Act: A Summary of the Law*, by Claudia Copeland.)

Subsequent to the enactment of the Ocean Dumping Act, DOD continued its efforts to seek safer methods for disposing of chemical weapons on the land, resulting in the development of incinerators to destroy them.[8] Although incineration remains the primary method by which chemical weapons are disposed of today, concerns about potential health and environmental risks from incineration have spurred the research and development of safer technologies to neutralize them. Under international agreement, the United States has committed to destroying its chemical weapons stockpile by 2012. However, there are questions as to whether this deadline can be met, considering the current capacity of existing disposal facilities. (For further discussion of ongoing efforts to dispose of chemical weapons at onshore facilities, see CRS Report RL32158, *Chemical Weapons Convention: Issues for Congress*, by Steve Bowman.)

POTENTIAL RISKS FROM DISPOSAL IN THE OCEAN

Although alternative technologies are now available for the onshore disposal of chemical weapons, DOD contends that, at the time, disposal in the ocean was deemed safer than disposal on land for two reasons. First, methods for disposal on land were initially limited to burial, presenting a long-term risk if weapons leaked or were recovered. Second, it was generally assumed that chemical agents released into seawater would be diluted to safe levels in the vastness of the ocean. However, this assumption was questioned in later years as scientific understanding of the effects of ocean pollution grew, leading to the general prohibition on the disposal of wastes into the ocean, including chemical weapons.

Exposure to chemical weapons can have numerous harmful effects on human beings. Depending on the particular chemical agent, these effects can include burns and sores on the skin, vomiting, respiratory dysfunction, mental impairment, damage to the immune and nervous systems, infertility, and death. Public health advocates have questioned whether possible exposure to such substances in seawater from leaking weapons may contribute to various symptoms experienced by coastal residents, swimmers, divers, fishermen, and individuals who may have consumed contaminated fish or shellfish. Marine conservationists and environmental advocates also have raised questions about the possible effects of chemical weapons agents on the marine environment, including the possible contribution to declines in populations of certain fish and other marine life in and around areas where weapons were dumped in the ocean.

The degree of risk from weapons leakingchemical agents into seawater depends on numerous factors. The extent to which an agent is diluted and the duration of exposure determine whether there is potential for harm. For example, most nerve agents are soluble and dissolve in water within several days. Less soluble agents still degrade over time as a result of hydrolysis. However, certain agents are less susceptible to hydrolysis, allowing them to remain in harmful forms for longer periods. For example, sulphur mustard in liquid or solid form turns into an encrusted gel when released in seawater. In this form, it can persist for many years before degrading.

Density is another critical factor. Chemical weapons agents denser than seawater tend to remain on the ocean floor, rather than float to shallower waters where they may present greater risk. For example, encrusted sulphur mustard is denser than seawater, making it unlikely to migrate off the ocean floor. However, ocean currents can disperse such substances along the seabed, spreading contamination beyond the location where the release occurred. Colder water temperatures also can slow degradation and allow contamination along the seabed to persist in harmful concentrations and forms for longer periods.

In addition to contamination of seawater, there have been concerns among the public that chemical weapons could wash ashore, or that they could be retrieved accidentally during dredging operations or trawl fishing along the seabed. The likelihood of such events is difficult to predict. Generally, the greater the depth of disposal, the less likely that accidental retrieval or washing ashore would occur. Although ocean currents could move weapons into shallower waters and present a greater safety risk, the accumulation of sediment and marine growth could help anchor weapons to the seabed, making them less susceptible to movement.

In its 2001 report, the Army documented few incidents of accidental retrieval of chemical weapons,[9] or of them washing ashore.[10] Only one other incident has been reported in the United States since then.[11] One could observe that these incidents are rare considering the thousands of chemical weapons dumped off the coasts of the United States over 50 years, and that the risks of such events appear *relatively* small. On the other hand, one could argue that even a few incidents prove that some risks do exist, and that study and monitoring of disposal areas are warranted to assess the likelihood of future risks.

RELEVANT SCIENTIFIC STUDIES IN EUROPE AND RUSSIA

Thus far, there have been no comprehensive scientific studies of potential risks to human health and the marine environment in specific areas of the ocean where chemical weapons were dumped off the coast of the United States. Therefore, it is difficult to provide definitive answers to questions about risks raised by public health and environmental advocates, marine conservationists, and the general public. However, there have been numerous scientific studies of risks from the disposal of chemical weapons off the coasts of Europe and Russia in the Atlantic Ocean, North Sea, Baltic Sea, and Mediterranean Sea. The former Soviet Union, Germany, Great Britain, and France disposed of chemicalweapons in these waters, including weapons captured during World War II. Although these studies focus on risks to coastal populations in Europe and Russia, and the marine environments of these waters, their findings may offer insights into potential risks from the disposal of chemical weapons off the coasts of the United States.

In 2005, the Imperial College of London compiled the findings of these European and Russian studies.[12] As a whole, they concluded that risks remain relatively small if weapons or persisting contamination on the seabed, such as encrusted sulphur mustard, remain undisturbed. On the other hand, human disturbances, such as dredging, trawl fishing, or work on underwater pipelines, caused risks to rise significantly. The studies linked instances of human exposure primarily to such disturbances, rather than to ocean currents washing weapons or contamination ashore.

In shallower waters, such as in the Baltic Sea, the studies noted that the greatest risk is to fishermen who reported many instances of catching encrusted sulphur mustard in their nets when trawling the seabed. However, the number of such instances declined as the depth of the water increased. Although the studies noted more instances of exposure than reported in the United States, the depths of disposal off the coasts of Europe and Russia generally are shallower than those off the coasts of the United States. The finding that potential risks appear to decrease relative to greater depths of disposal suggests that there may be less potential risk domestically than in the waters of these foreign nations.

RESPONSE OPTIONS AND ISSUES

As DOD has disclosed more information about the past disposal of chemical weapons in the ocean, interest in how best to respond to potential risks has grown among the public. Thus far, the U.S. Army has prepared materials for commercial maritime industries to educate individuals about the hazards of chemical and conventional weapons that may be present in the ocean. These materials include safety guidelines in the event that weapons are accidentally retrieved from the seabed, and provide contact information to inform federal officials of the presence of such weapons, so that appropriate actions may be taken to ensure public safety. However, the locations of the disposal sites are not disclosed, preventing the public from being able to avoid areas where they may encounter weapons dumped offshore.

The primary obstacle to responding to potential risks is locating the weapons in the ocean. The lack of coordinates for most of the disposal sites, and the possibility that ocean currents may have moved weapons beyond these areas, makes finding the weapons difficult at best, if not impracticable in some cases. In the event that the weapons are located, one option to prevent exposure would be to leave the weapons in place and warn the public to avoid these areas. This option may address immediate risks and avoid new risks associated with retrieving weapons from the ocean (see below). However, the long-term risks of leaving the weapons in place is uncertain, because of a lack of information on the leaking of chemical agents in the ocean and the extent to which ocean currents may cause munitions to migrate to shallower waters or wash ashore in the future. Leaving the weapons in place and publicly disclosing their location also could present risks to national security in the event that individuals may retrieve weapons and use them for harmful purposes.

If the weapons are found, a more difficult option would be to remove them from the seabed and to remediate contaminated seawater surrounding these areas. Once retrieved, these weapons would need to be disposed of safely onshore, through incineration or emerging alternative technologies to neutralize them. Although this option may more effectively address long-term risks, it likely would require substantial financial resources and time, and would be technically challenging. This option also could present new risks, such as risks to workers who would remove and transport the weapons, and to populated areas in the event of an accidental release during transporting or disposing of the weapons at an onshore facility. Regardless of the desired option, the practical feasibility of responding to potential risks is limited by the many inherent challenges discussed below.

Locating Disposal Sites

Locating where chemical weapons were dumped in the ocean would be the first step toward assessing potential risks and determining the feasibility of response actions. However, the *precise* coordinates of most of these sites are unknown, both in U.S. and foreign waters. The specificity of the geographic locations of the disposal sites in historical records varies widely. Rough coordinates are known or assumed for some sites, whereas only approximate distances from the coast are identified for others. The location of many sites is completely unknown, with only the body of water identified, such as the Atlantic or Pacific Ocean in general. Consequently, attempts to locate most of the disposal sites would be difficult at best, if not impracticable in some cases. The Army's 2001 report indicated that only four disposal sites where the exact location is known have been surveyed, with no contamination detected at the time. The last survey occurred in 1975. Some have advocated that at least these four known sites be revisited to determine if chemical weapons agents may have leaked since then.

Identifying Weapons and Containers

In the event that a disposal site is located, identifying *individual* weapons underwater still is likely to be difficult. At many sites, weapons were disposed of one-by-one, or "loose dumped" as the Army describes in its 2001 report. At these sites, ocean currents may have moved weapons and containers far enough away from their original point of disposal to make finding them problematic. Weapons also may be difficult to locate, depending on their size and depth, because of the accumulation of sediment and biological growth over many years. Sites where there are topographical irregularities in the ocean floor also could be challenging to survey with sonar technologies, because of the difficulty in distinguishing among natural formations that may be similar to weapons in size and shape. The Army's report listed numerous sites where weapons were sunk inside the hulls of obsolete vessels, which may be easier to identify because of their size and shape. As noted above, DOD has located and surveyed four of sites where weapons were sunk inside obsolete vessels, with the last surveyed in 1975.

Retrieving Munitions for Onshore Disposal

Safely retrieving weapons from the ocean floor is generally more challenging the greater the depth. The suspected disposal depth varies widely. In its 2001 report, the Army indicated disposal depths ranging from 200 feet in the Gulf of Mexico to over 16,000 feet off the coast of Florida. However, these records do not indicate the depths of many disposal sites. Another factor is structural integrity. Weapons casings and containers may be weakened as result of water pressure, and the corrosive effects of saltwater, over prolonged periods. Especially at greater depths, changes in water pressure during retrieval could cause weakened casings and containers to rupture, resulting in the release of chemical weapons agents into shallower waters where they may pose greater risks of exposure. Existing leaks also could be exacerbated in such circumstances, increasing the quantity of chemical weapons agents released.

Remediating Contaminated Seawater

Experience in remediating contaminated seawater primarily has been limited to the removal of oil and fuel spilled on surface waters. Oil and fuel tend not to mix with water because of their physical composition, making them easier to remove. Chemical weapons agents that are soluble, such as nerve agents, mix easily with seawater, making removal impracticable.

However, other agents that do not mix well with water, such as sulphur mustard, could be easier to remove. The depth at which contamination occurs is another important factor. The feasibility of remediating contamination far below the surface is highly uncertain because of the lack of experience in performing this type of remediation.

Although there may be a greater possibility for effective remediation if insoluble chemical weapons agents were to migrate to the surface, currents and winds in open waters could spread such agents rapidly over unmanageably large areas, especially at sites far from the coast where currents and winds can be stronger. The Army's 2001 report indicated at least one instance of remediation of contaminated surface waters during World War II in Bari Harbor, Italy, where a German air raid destroyed a U.S. Naval vessel containing mustard gas bombs and resulted in contamination of the harbor. However, the Army's report does not provide any information to assess the degree or adequacy of the cleanup, relative to today's standards.

Costs

Considering that much is unknown about the quantity and condition of weapons dumped in the ocean, developing reliable estimates of the costs to respond to potential risks is, and will continue to be, nearly impossible without such information. The availability of federal funding to pay the costs of response actions is an issue as well, whatever amount they may be. There already is ongoing debate within Congress about the adequacy of funding to pay for the cleanup of land-based contamination on military lands, for which the estimated costs are substantial.

In its most recent environmental report to Congress in March 2006, DOD estimated that a total of $34.4 billion would be needed to complete cleanup at active military installations, closed bases, and other former military properties.[13] Certain factors could cause these costs to rise, including efforts to clean up unexploded ordnance (UXO), the cleanup of another round of base closings approved in 2005 to prepare these properties for civilian reuse, and possibly more stringent cleanup standards in the future.

Addressing the disposal of chemical weapons in the ocean could add to these costs, placing further demands on military funding for cleanup. Appropriating additional funds to meet these needs could prove challenging, considering the many other competing needs within the overall federal budget, such as other military needs, homeland security, and domestic programs, and constraints on spending as a result of the federal budget deficit.

RESPONSE AUTHORITIES

Although the Ocean Dumping Act has specifically prohibited the disposal of chemical weapons in the ocean since 1972, it does not explicitly authorize response to human health and environmental risks resulting from past disposal. However, there are at least three federal statutes that one might examine for possible authority to respond to risks from the past disposal of chemical weapons in certain areas of the ocean off U.S. shores. These statutes include the Comprehensive Environmental Response, Compensation, and Liability Act (CERCLA),[14] the Clean Water Act,[15] and the Resource Conservation and Recovery Act (RCRA).[16] Of these three statutes, CERCLA appears to raise the fewest questions regarding applicability, because it explicitly applies within specific ocean waters off the coasts of the United States. RCRA appears to raise the most questions of the three, because the applicability of that statute to ocean waters is unclear.

The applicability of response authorities in federal law hinges upon whether a particular substance is *legally* defined as a "hazardous substance" or "pollutant or contaminant," in the case of CERCLA; a "hazardous substance" or "pollutant," in the case of the Clean Water Act; or a "solid waste" or "hazardous waste," in the case of RCRA. Response authorities in these statutes also hinge upon whether there is an "imminent and substantial" danger to public health or welfare, or the environment. Although what constitutes imminent and substantial in this context is not defined in federal statute, courts generally have deferred to the Environmental Protection Agency (EPA) in making such determinations.

Further, the scope of these three statutes has not been interpreted to apply outside the United States and the adjacent waters over which the United States asserts limited jurisdiction (explained below). The Army's acknowledgment that the U.S. Armed Forces disposed of chemical weapons off the coasts of foreign nations raises questions regarding the liabilityof the United States under international law, or under the domestic laws of such foreign nations in the event that disposal occurred within their respective territorial seas.

There appears to be existing authority in federal law for the United States to respond to potential risks from chemical weapons disposed of off its shores.

However, as discussed above, the primary obstacle to responding to potential risks is locating the weapons in the ocean. In the event that the weapons are found, many factors, such as technical capabilities, the possible introduction of new risks, and costs, could constrain the types of actions that could be carried out under these authorities. Possible applicability of each authority is discussed further below.

CERCLA

Section 104(a) of CERCLA authorizes the President to respond to a release, or a threat of a release, of a hazardous substance, pollutant, or contaminant into the environment.[17] The authority to respond to a release of a pollutant or contaminant is dependent upon whether there is an imminent and substantial danger to public health or welfare. The statute does not explicitly require such danger to be present to respond to a release or threatened release of a hazardous substance, primarily because the *nature* or *characteristics* of a hazardous substance clearly imply such danger by the fact that such substance is "hazardous." In practice, the President typically delegates the authority to respond to a release or threatened release to one or more federal agencies.

Authorized response actions may include removal or remediation or both. A "removal" is defined more broadly in CERCLA than the literal removal of contamination or source of contamination. Rather, a removal is an immediate or short-term response to an exposure threat, including but not limited to containing waste, preventing access to contaminated areas, and providing emergency assistance.[18] A "remedial action" may include many of the same activities as a removal, but is meant as a permanent remedy.[19] Remedial actions are also subject to more thorough review prior to implementation, including opportunity for public comment. In practice, a removal action is often performed as an *interim* response to animmediatethreat, while a remedial action is planned to provideprotection over the *long-term*.

Although CERCLA does not explicitly address threats from the release of chemical warfare agents, a "hazardous substance" is defined in that statute[20] to include those substances defined as such by EPA under the Clean Water Act.[21] EPA has listed many chemical warfare agents as hazardous substances under this authority.[22] "Pollutant or contaminant" also is broadly defined in CERCLA in a way that arguably could include all chemical warfare agents, because of the physiological effects on humans and nonhuman organisms that can result from exposure.[23]

Further, "release" is defined in CERCLA to include dumping, disposing, or leaking into the environment, including the discarding of containers, whether or not they have leaked.[24] The "environment" is defined to include the ocean waters of the U.S. Exclusive Economic Zone (EEZ).[25] The outer boundary of the EEZ is 200 nautical miles seaward of the baseline from which the territorial sea is measured, which is no greater than 12 nautical miles from U.S. shores.

Thus, the past disposal of chemical weapons by the U.S. Armed Forces in the ocean within the seaward boundary of the U.S. EEZ arguably could be considered a release of a hazardous substance, pollutant, or contaminant into the environment,[26] thereby authorizing the federal government to take response actions to protect public health or welfare, or the environment. Although CERCLA could be interpreted as being applicable, finding the weapons in the ocean is a significant obstacle to taking a response action. If the weapons are found, many factors still could constrain the feasibility of certain actions, as explained above.

Clean Water Act

Section 311(c) of the Clean Water Act authorizes the President to "ensure effective and immediate removal of a discharge, and mitigation or prevention of a substantial threat of a discharge, of oil or a hazardous substance" into the navigable waters and adjoining shorelines of the United States, and into the ocean waters of the U.S. EEZ.[27] As is the case with CERCLA, the President typically delegates this responsibility to one or more federal agencies.

As noted above, EPA has categorized many chemical warfare agents as hazardous substances both for the purposes of the Clean Water Act, and in turn CERCLA. Discharge is defined in Section 311(a)(2) of the Clean Water Act to include "dumping" and "leaking."[28] Depending on how these statutory definitions are interpreted, Section 311(c) of the Clean Water Act could authorize the President to remove chemical weapons from the ocean within the seaward boundary of the U.S. EEZ, including those that may be leaking chemical agents into seawater. Similar to CERCLA, the applicability of the Clean Water Act does not necessarily mean that a desired response action could be taken in practice. As indicated above, finding the weapons is a significant obstacle, and if the weapons are found, many factors could constrain the feasibility of response actions.

Section 504 of the Clean Water Act also grants "emergency powers" to EPA to bring suit to compel persons contributing to pollution, or to take other action, to stop the discharge of pollutants that present an imminent and substantial endangerment to human health or welfare.[29] "Pollutant" is defined in the Clean Water Act for the purposes of this section to include munitions and chemical wastes.[30] However, there are longstanding questions regarding the ability of one federal agency to enforce an action against another, in the case of chemical weapons, likely EPA enforcing an action to be taken by DOD. There also are questions as to whether Section 504 allows EPA to order a response be taken, in the event that the original action that caused the pollution is no longer ongoing but has long since ended, such as the disposal of chemical weapons in the ocean from the World War I era through 1970.

RCRA

The applicability of response authorities in RCRA to risks from chemical weapons in the ocean is less clear than in CERCLA and the Clean Water Act. Section 7003 of RCRA authorizes EPA to take any action necessary to respond to an imminent and substantial endangerment to human health or the environment resulting from the disposal of a solid waste, including a hazardous waste.[31] As defined in RCRA, disposal includes "dumping" into water, but does not specify whether ocean waters are included.[32] A "solid waste,"[33] or a "hazardous waste"[34] as a type of solid waste, also are not explicitly defined in RCRA to include chemical weapons or agents.

However, RCRA does authorize EPA to promulgate regulations to determine when munitions, including chemical munitions, are considered a solid waste, or a hazardous waste, for purposes of that statute.[35] EPA promulgated such regulations in 1997.[36] Among other circumstances, the regulations specify that a military munition, including a chemical munition, is considered a solid waste, and by its characteristics a hazardous waste, when it is abandoned as a result of disposal. However, like the statute, the regulations do not clarify their applicability to disposal in ocean waters.

If these regulations were interpreted to apply to ocean waters, and the disposal of chemical munitions resulted in imminent and substantial endangerment to human health or the environment, EPA arguably could have response authorities under Section 7003 of RCRA. EPA also has authority under Section 6001(b) of RCRA to issue an administrative order to another federal agency, likely DOD in the case of chemical weapons, to specify response actions that EPA deems necessary.[37] Section 6001(a) of RCRA specifies that federal agencies must comply with such orders.[38] Similar to CERCLA and the Clean Water Act, whether a desired response action could be taken would depend on whether the weapons could be found, and whether the desired action would be practically feasible.

LEGISLATION ENACTED IN THE 109TH CONGRESS

In its second session, the 109th Congress enacted legislation to gain a better understanding of the past disposal of chemical weapons in the ocean by the U.S. Armed Forces, and the potential health, safety, and environmental risks that these weapons may pose. Section 314 of the John Warner National Defense Authorization Act for FY2007 (P.L. 109-364, H.R. 5122) requires further review of historical records to attempt to identify areas where the U.S. Armed Forces disposed of both chemical *and* conventional weapons off U.S. shores. Although the past dumping of chemical weapons motivated initial concerns, there also were questions about the extent to which the U.S. Armed Forces may have dumped surplus or damaged conventional weapons into the ocean. To learn more about potential risks, Section 314 also requires the Secretary of Defense to research the effects of these weapons on the ocean environment and to monitor identified sites to determine whether contamination is *currently* being released or significant health or safety risks are present.

The provisions in Section 314 were based on legislation introduced earlier in the second session of the 109th Congress in the Hawaiian Waters Chemical Munitions Safety Act of 2006 (H.R. 4778) and its companion bill in the Senate (S. 2295). Whereas these bills focused on chemical weapons dumped off the coast of Hawaii, the scope of Section 314 in P.L. 109-364 was expanded to encompass all U.S. coastal areas and to include conventional weapons, as noted above. An analysis of provisions in Section 314 as enacted, and a comparison to the original provisions in the House and Senate versions of H.R. 5122, are provided below.

Subsection(a) of Section 314 requires the Secretary of Defense to review historical records to determine the number and "probable" locations of sites where the U.S. Armed Forces disposed of military munitions within U.S. coastal waters, the size of these sites, and the types of munitions dumped at these locations. The original House bill also would have required the disclosure of the *quantities* of munitions dumped at each site, whereas the enacted bill did not. The areas of the ocean covered in the historical review are to include areas extending from the U.S. shoreline to the outer boundary of the Outer Continental Shelf.[39] The Secretary of Defense is required to *request* the assistance of the U.S. Coast Guard, the National Oceanic and Atmospheric Administration (NOAA), and other relevant federal agencies in reviewing historical records of disposal in the ocean. This provision is consistent with the original House bill, whereas the original Senate bill would have *required* the cooperation of these agencies.

Subsection(a) also requires the Secretary of Defense to release the information compiled from its historical review on an annual basis. The Secretary is required to include this information in the Department's annual report to Congress on its environmental restoration activities.[40] However, the Secretary is authorized to withhold from the public the "exact nature and locations of munitions," if he determines that the potential for unauthorized retrieval of these weapons could present a significant threat to national defense or public safety. Determining when it would be appropriate to withhold information about an individual site would be at the discretion of the Secretary.

Subsection(a) further requires the Secretary of Defense to complete the historical review of munitions disposal sites within a time frame necessary for it to be included in the Department's FY2009 environmental restoration report to Congress. The original House bill did not specify a time frame for the completion of the historical review. Instead, it would have required annual reporting as the Secretary compiles records of munitions disposal in the ocean, with an implied presumption that the review would be complete at some point in time.

Subsection(b) requires the Secretary of Defense to share information on the disposal of munitions in the ocean with the Secretary of Commerce to assist NOAA in preparing nautical charts and other navigational aids to identify known or likely hazards to the public, including commercial shipping and fishing operations. This subsection also requires the Secretary of Defense to continue activities to inform the public of the possible hazards of coming into contact with military munitions on the seabed, and to continue efforts to identify appropriate actions to mitigate such hazards if contact does occur. As noted earlier, the U.S. Army has prepared materials for commercial maritime industries to educate individuals about the hazards of munitions that may be present in the ocean. These materials include safety guidelines in the event that munitions are accidentally retrieved from the seabed, and provide contact information to inform federal officials of the presence of munitions, so that appropriate actions may be taken to ensure public safety.

Subsection(c) requires the Secretary of Defense to research the effects of munitions disposed of in coastal waters on the ocean environment and those who "use" ocean waters. The scope of "effects" is not specified. Presumably, examined effects could include human health, safety, and environmental risks, and the economic impacts of potential damage to marine resources. However, the scope could be narrower or broader than these potential effects, and presumably would be at the discretion of the Secretary. Subsection(c) also authorizes the Secretary to award grants and enter into cooperative agreements to "qualified" entities to perform this research, but does not stipulate criteria for determining whether an entity would be qualified for this task. This determination therefore presumably would be at the discretion of the Secretary, whereas the original House bill explicitly stated that this determination would have been at the discretion of the Secretary.

To conduct this research, the Secretary is required to select at least two "representative" (i.e., typical) sites along the Atlantic coast, two along the Pacific coast (including the coast of Alaska), and two off the coast of Hawaii. Factors for selecting representative sites include depth, water temperature, nature of the military munitions, and proximity to coastal populations. The physical scope of the study of disposal sites is ambiguous in terms of the surface area and volume of seawater that is to be examined. Required research at each site is to entail (1) sampling and analysis of ocean waters and the seabed at or adjacent to the locations where munitions were dumped; (2) assessment of the long-term effects of exposure to seawater on military munitions, particularly *chemical* munitions; and (3) development of "effective" safety measures when dealing with (i.e., handling) military munitions that have been disposed of in seawater.

Subsection(c) further requires a study ofthe feasibility of removing or otherwise remediating munitions in the ocean and requires study of the impacts on the ocean environment and those who use it, including public health risks. The original House bill did not contain similar language regarding study of the feasibility of removal or remediation, but did include language that would have required an epidemiological study of the effects of munitions in the ocean on human health in coastal populations located in the "vicinity" of disposal sites. However, it is unclear how close a community would had to have been located to an offshore disposal site for it to have been included in such study.

If contamination *currently* is being released into ocean waters at identified sites or a significant health or safety risk is present, Subsection(d) requires the Secretary of Defense to put "appropriate" mechanisms into place at those sites to monitor the contamination. However, such mechanisms are not specified in the law or in the accompanying conference report on the final version of the bill. If monitoring would be so required, the Secretary must report to the congressional defense committees on additional measures that may be needed to address the release of contamination or potential risks. Although monitoring information may be used to inform decisions as to whether further response is needed to address potential risks, neither the final bill signed into law nor the original House or Senate versions require or authorize such action. As explained earlier, at least three existing federal statutes could be interpreted to provide authority for responding to such risks, without possibly requiring new authority from Congress.

As the House and Senate originally proposed, Section 314 of P.L. 109-364 does not authorize a specific amount of funding to carry out the historical review, research, and monitoring of munitions disposal sites in the ocean. However, the law provides general authority[41] for the Secretary of Defense and other relevant federal agencies to *engage* in these activities. Whether or not a specific funding level is authorized for an activity, actual funding to pay for it is subject to appropriations. In the 109th Congress, neither of the two FY2007 defense appropriations bills (H.R. 5631, as signed into law, or H.R. 5385, as passed by the House and Senate) contained specific funding to implement a study of offshore munitions disposal sites.

Absent a set-aside appropriation, DOD still could allocate funds for a study of offshore munitions disposal sites in FY2007 out of funds appropriated to accounts for broader purposes which are within its discretion. Considering that the study of offshore sites would involve the assessment of contamination, DOD's environmental restoration accounts may be a likely source of funding. However, as discussed earlier, there are substantial funding needs estimated for the cleanup of land-based contamination with these funds, raising questions about the availability of funding within these accounts to address offshore contamination. The many other competing national security needs within DOD's budget also could limit available funding, unless Congress were to provide targeted appropriations for the study.

The funds needed to study offshore munitions disposal sites is uncertain. The cost of reviewing historical records to identify disposal sites would likely be relatively small, primarily involving personnel expenses. However, the cost of researching and monitoring identified sites could be substantially higher, involving the use of vessels to reach offshore areas, scientific equipment to gather seawater samples at possibly great depths, specialized personnel trained in the operation of such equipment, and laboratory analysis of monitoring data. Research and monitoring costs would depend on numerous factors, including the geographic scope of the sites, their distance from the shore, the depth at which munitions are present, and the methodologies used to evaluate contaminants and associated risks.

CONCLUSION

Although the Army has disclosed more information than previously available, much remains unknown about the exact quantities, types, and present locations of chemical and conventional weapons that the U.S. Armed Forces dumped in the ocean. Incomplete historical records significantly limit the ability to identify and assess the condition of these weapons, particularly to determine whether chemical agents may have leaked, or are likely to do so. Section 314 of P.L. 109-364 (H.R. 5122) requires the Secretary of Defense to attempt to identify sites where weapons were dumped off U.S. shores, and to research and monitor identified sites where contamination is being released or there is a significant health or safety risk. However, considering the lack of complete records of disposal released thus far, it would appear that this task would be difficult at best, if not impracticable in some cases, especially in instances in which the known location of disposal is in a broad area of the ocean in very deep waters.

Assessing the degree of potential risks is nearly impossible without knowing the specific location and condition of weapons on the seabed. Risks to human health and the environment could be lessened if the volume and effects of seawater may have diluted leaked chemical agents into less threatening concentrations or degraded them into less dangerous components. However, the possibility of harmful exposure remains, especially in the event that a weapon is accidentally retrieved or washes ashore, and a chemical agent is released in a harmful concentration and form or a “live” conventional weapon were to detonate. While the number of such instances has been rare relative to the thousands of weapons dumped in the ocean, the possibility of future instances and associated risks remain.

Retrieving weapons from the ocean to address potential risks is fraught with many practical challenges. The primary obstacle is locating the weapons in the ocean. In the event that the weapons are found, retrieving them would introduce new risks if weapons casings ruptured in the ascent from great depths, or if an accidental release occurred during transport to onshore facilities for disposal. Considering these challenges and risks, leaving identified weapons in place, and warning the public to avoid areas of the ocean where they were dumped, may be more feasible. However, the long-term risks of leaving located weapons in place is uncertain due to a lack of information on the effects of chemical agents in ocean waters, and the extent to which weapons may migrate along the seabed to shallower waters or wash ashore over time where they may present greater risks. Further, public disclosure of the location and types of weapons could present national security risks, in the event that individuals were to retrieve these weapons and use them for harmful purposes.

Regardless of which option is desired, finding most of the weapons in the ocean to respond to potential risks would appear to be highly unlikely. The substantial challenge of accurately identifying the boundaries of former disposal areas without complete historical records, and the possibility that ocean currents may have moved weapons and contamination beyond these boundaries, makes the implementation of any response option difficult at best, if not impracticable in some cases. The feasibility of responding to potential risks is further complicated by unknown costs of response actions and the uncertain availability of federal funding to pay for them.

End Notes

[1] Department of Defense. U.S. Army Research, Development, and Engineering Command, Aberdeen Proving Ground, Maryland. Corporate Information Office. Historical Research and Response Team. *Off-shore Disposal of Chemical Agents and Weapons Conducted by the United States*. March 29, 2001. 15 pp.

[2] According to the Army's 2001 report, chemical weapons disposed of in the ocean by the U.S. Armed Forces included surplus and damaged bombs, rockets, projectiles, and other munitions containing chemical warfare agents, and barrels, cylinders, and other containers filled with surplus chemical warfare agents produced for use in munitions. In this report, the term *chemical weapons* includes of all these items and, as such, refers to both chemical munitions and containers of chemical warfare agents.

[3] Sulphur mustard is a viscous liquid that becomes solid at 58 degrees Fahrenheit. Chemical weapons contain sulphur mustard in either form depending on temperature. Upon impact, chemical weapons release sulphur mustard in a gaseous vapor, referred to as "mustard gas."

[4] Common nerve agents include tabun, sarin, soman, and VX. They are liquid in form, but when exposed to the air, they evaporate quickly into a gas. Chemical weapons contain nerve agents in liquid form, to be released as a gas upon delivery to an intended target.

[5] National Academy of Sciences. *Disposal Hazards of Certain Chemical Warfare Agents and Munitions*. June 24, 1969. 14 pp.

[6] 33 U.S.C. 1401 et seq. The "Ocean Dumping Act" is the common reference to Title I of the Marine Protection, Research, and Sanctuaries Act of 1972 (MPRSA, P.L. 92-532), as amended.

[7] 33 U.S.C. 1412.

[8] The U.S. Army Chemical Materials Agency administers the disposal of chemical munitions, including the operation of four disposal facilities located in Anniston, Alabama; Pine Bluff, Arkansas; Umatilla, Oregon; and Tooele, Utah. For further information, see the agency's website at [http://www-pmcd.apgea.army.mil].

[9] For example, a fisherman accidentally retrieved a container of sulphur mustard off the coast of Australia in 1970 that the U.S. Armed Forces had dumped in 1945 along with thousands of tons of chemical weapons. In 1976, a dredging operation off the coast of Hawaii accidentally retrieved a mortar round containing chemical agents, injuring one crewman. The Army suspects that the mortar round was one of thousands of tons of chemical weapons dumped in 1944 off Pearl Harbor.

[10] For example, a mustard gas bomb floated ashore in the Gulf of Mexico in 1946 (location unspecified) after it and 32 others were disposed of 20 miles off the coast at depths ranging from 200 to 600 feet. The bomb was recovered safely. Also in 1946, an unspecified number of mustard gas bombs appeared floating in Manfredonia Bay, Italy, within three weeks after they had been disposed of 54 miles from shore. In the 1970s (year not specified), a container of sulphur mustard washed ashore off the coast of Australia in the same area where a fisherman accidentally retrieved a container in 1970. According to the Army, the government of Australia has since designated this area hazardous.

[11] A report by U.S. Army personnel acknowledged the accidental retrieval of a World War I era mustard gas munition by a clam dredging operation off the coast of New Jersey in 2004. The report, *Mitigating the Possible Damaging Effects of Twentieth-Century Ocean Dumping of Chemical Munitions*, by Emily E. Baine and Margaret P. Simmons, was prepared independently by these authors, not by the U.S. Army. Numerous press reports also have acknowledged this incident.

[12] Imperial College of London. *Munitions Dumped at Sea: A Literature Review*. June 2005. 90 pp. As noted in the Army's 2001 report, the U.S. Armed Forces also disposed of chemical weapons off the coasts of Europe in the Atlantic Ocean, North Sea, and Mediterranean Sea. The studies reviewed in the Imperial College report do not distinguish between risks from U.S. disposal and disposal by other nations in these foreign waters.

[13] Department of Defense. *Defense Environmental Programs Annual Report to Congress for Fiscal Year 2005*. March 2006. Appendix J, p. J-6-1.

[14] 42 U.S.C. 9601 et seq.

[15] 42 U.S.C. 1251 et seq.

[16] 42 U.S.C. 6901 et seq.

[17] 42 U.S.C. 9604(a).

[18] 42 U.S.C. 9601(23).

[19] 42 U.S.C. 9601(24).

[20] 42 U.S.C. 9601(14).

[21] 33 U.S.C. 1321(b)(2)(A).

[22] 40 C.F.R. 116.4.

[23] 42 U.S.C. 9601(33). CERCLA defines "pollutant or contaminant" as "any element, substance, compound, or mixture, including disease-causing agents, which after release intothe environment and upon exposure, ingestion, inhalation, or assimilation into any organism,either directly from the environment or indirectly by ingestion through food chains, will ormay reasonably be anticipated to cause death, disease, behavioral abnormalities, cancer,genetic mutation, physiological malfunctions (including malfunctions in reproduction) orphysical deformations, in such organisms or their offspring...."

[24] 42 U.S.C. 9601(22). "Release" includes dumping, disposing, or leaking into the environment, including the "abandonment or discarding of barrels, containers, and other closed receptacles containing any hazardous substance or pollutant or contaminant." Thus, the dumping or disposal of containers of chemical agents could constitute a release, and munitions containing chemical agents could constitute a *closed receptacle*, and the dumping or disposing of them therefore a release.

[25] 42 U.S.C. 9601(8). In this subsection, the "ocean waters" of the United States specifically refer to ocean waters of which the natural resources are under the exclusive management authority of the United States under the Magnuson-Stevens Fishery Conservation and Management Act (16 U.S.C. 1801 et seq.), which extend to the seaward boundary of the U.S. EEZ.

[26] Liability under CERCLA is retroactive. Therefore, the fact that the U.S. Armed Forces disposed of chemical weapons in the ocean long before 1980 when Congress enacted CERCLA does not remove the disposal of these weapons from the applicability of that statute.

[27] 33 U.S.C. 1321(c).

[28] 33 U.S.C. 1321(a)(2).

[29] 33 U.S.C. 1364(a).

[30] 33 U.S.C. 1362(6).

[31] 42 U.S.C. 6973(a).

[32] 42 U.S.C. 6903(3).

[33] 42 U.S.C. 6903(27).

[34] 42 U.S.C. 6903(5).

[35] 42 U.S.C. 6924(y).

[36] 40 C.F.R. 266.200.

[37] 42 U.S.C. 6961(b).

[38] 42 U.S.C. 6961(a).

[39] Section 314 of P.L. 109-364 uses the existing statutory definition of Outer Continental Shelf referenced in the Outer Continental Shelf Lands Act (43 U.S.C. 1331).

[40] 10 U.S.C. 2706(a) requires the Secretary of Defense to submit an annual report to Congress on its environmental restoration activities, including the number of contaminated sites, the status of cleanup by site, incurred costs, and estimates of future costs to complete cleanup. 10 U.S.C. 2706(b) requires the Secretary of Defense to submit an annual report to Congress on other environmental activities on military lands, including compliance with pollution control requirements, pollution prevention efforts, and compliance with natural resource and historic preservation requirements. Beginning in FY2004, DOD consolidated these two annual environmental reports into one document covering all environmental programs and activities. Accordingly, the information required in Section 314 of P.L. 109-364 presumably would be included in this consolidated annual environmental report.

[41] Statutory authority for an agency to carry out an action constitutes what is often referred to as "program authority." Inauthorizingan activity, Congress often also provides "funding authority" that may specify a dollar amount for certain fiscal years, or maysimplyauthorize "such sums as may be necessary." Authorized funding levels are non-binding and are subject to annual appropriations. As such, Congress may appropriate amounts that differ from a funding authorization. If funding authority is not provided or has expired, Congress may still appropriate funding for an activity for which it has provided program authority. A department or an agency also may fund an activity for which it has program authority even if Congress does not explicitly set aside funding for it, if appropriated funds are available to allocate to that activity at the discretion of the department or agency.

CHAPTER SOURCES

The following chapters have been previously published:

Chapter 1 – These remarks were delivered as Written Testimony of Richard A. Feely, Ph.D., Pacific Marine Environmental Laboratory, National Oceanic and Atmospheric Administration, U.S. Department of Commerce, before the Committee on Commerce, Science, and Transportation, Subcommittee on Oceans, Atmosphere, Fisheries and Coast Guard, U.S. Senate, dated May 10, 2007.

Chapter 2 – This is an edited, excerpted and augmented edition of a United States Congressional Research Service publication, Report Order Code 98-869, dated August 13, 2007.

Chapter 3 – This is an edited, excerpted and augmented edition of a United States Congressional Research Service publication, Report Order Code R40143, dated January 16, 2009.

Chapter 4 - This is an edited, excerpted and augmented edition of a United States Congressional Research Service publication, Report Order Code RS20028, dated November 17, 2008.

Chapter 5 – This is an edited, excerpted and augmented edition of a United States Congressional Research Service publication, Report Order Code RL33705, dated September 2, 2008.

Chapter 6 - This is an edited, excerpted and augmented edition of a United States Congressional Research Service publication, Report Order Code RL33432, dated Januart 3, 2007.

INDEX

A

B

C

D

E

F

G

H

I

J

K

L

M

N

O

P

Q

R

S

T

U

V

W

Y

Z